SpringerBriefs in Philosophy

SpringerBriefs present concise summaries of cutting-edge research and practical applications across a wide spectrum of fields. Featuring compact volumes of 50 to 125 pages, the series covers a range of content from professional to academic. Typical topics might include:

- A timely report of state-of-the art analytical techniques
- A bridge between new research results, as published in journal articles, and a contextual literature review
- A snapshot of a hot or emerging topic
- An in-depth case study or clinical example
- A presentation of core concepts that students must understand in order to make independent contributions

SpringerBriefs in Philosophy cover a broad range of philosophical fields including: Philosophy of Science, Logic, Non-Western Thinking and Western Philosophy. We also consider biographies, full or partial, of key thinkers and pioneers.

SpringerBriefs are characterized by fast, global electronic dissemination, standard publishing contracts, standardized manuscript preparation and formatting guidelines, and expedited production schedules. Both solicited and unsolicited manuscripts are considered for publication in the SpringerBriefs in Philosophy series. Potential authors are warmly invited to complete and submit the Briefs Author Proposal form. All projects will be submitted to editorial review by external advisors.

SpringerBriefs are characterized by expedited production schedules with the aim for publication 8 to 12 weeks after acceptance and fast, global electronic dissemination through our online platform SpringerLink. The standard concise author contracts guarantee that

- an individual ISBN is assigned to each manuscript
- each manuscript is copyrighted in the name of the author
- the author retains the right to post the pre-publication version on his/her website or that of his/her institution.

Miguel Fuentes

Complexity, Emergence and the Evolution of Scientific Theories: Towards a Predictive Epistemology

Springer

Miguel Fuentes
Santa Fe Institute
Santa Fe, NM, USA

IIF—SADAF
Buenos Aires, Argentina

Instituto de Sistemas Complejos de Valparaíso
Valparaíso, Chile

ISSN 2211-4548 ISSN 2211-4556 (electronic)
SpringerBriefs in Philosophy
ISBN 978-3-031-99993-2 ISBN 978-3-031-99994-9 (eBook)
https://doi.org/10.1007/978-3-031-99994-9

This Springer imprint is published by the registered company Springer Nature Switzerland AG
The registered company address is: Gewerbestrasse 11, 6330 Cham, Switzerland

Acknowledgements

I would like to sincerely thank my colleagues at the Santa Fe Institute, who have engaged with me, at times through intense debate, on the ideas developed in this book. I hold a special memory of Murray Gell-Mann, who generously shared some of his thoughts with me at SFI, surrounded by the remarkable landscapes of New Mexico. I am also thankful to Professor Hernán Miguel, mentor and friend from the Argentine Society for Philosophical Analysis, whose guidance has been truly valuable. Finally, my thanks go to Beatriz Balmaceda and Tara Amelia for their steady support and constant presence throughout this journey.

"Again the pressure began, and at 5 p.m. I ordered all hands on to the ice."
Shackleton abandoning the Endurance, trapped in pack ice in 1915 at 69° 05′S, 51° 30′W.

Competing Interests The author has no competing interests to declare that are relevant to the content of this manuscript.

Introduction

The concept of emergence refers to phenomena that exhibit novel characteristics arising from more fundamental interactions yet retain a certain degree of autonomy from the processes that initially generate them. This notion is widely applied across disciplines. In physics, it underpins the study of phase transitions (Bedau & Humphreys, 2008; Landau & Lifshitz, 1994) and symmetry-breaking (Nicolis, 1995). In biology, life itself is often cited as an emergent phenomenon (Schrödinger, 1944); in cognitive science, consciousness represents another paradigmatic case (Paster, 2006); and in the social sciences, emergent properties emerge from individual interactions, giving rise to collective behavior (Helbing, 2010; Sayer, 2010). These various discussions converge on a central issue in philosophy: the nature of emergent properties—an enduring question that has gained renewed importance through recent advances in multiple scientific fields (Bedau & Humphreys, 2008; Humphreys, 2009). This book re-examines the problem of emergence through the lens of contemporary developments, particularly those arising from physics, information theory, and complexity science—areas that are often studied in isolation. A central theme of our inquiry is the relationship between emergent phenomena and both the structural properties of systems and the epistemological limits and transformations of the theories we use to understand them. In this context, we introduce the concept of predictive epistemology: the idea that increasing theoretical complexity—especially when required to maintain internal coherence in the face of anomalies and new data—may signal the approach of a paradigm shift. This perspective offers a novel way of anticipating the evolution of scientific frameworks by analyzing the dynamics of complexity itself.

The structure of the book reflects this interdisciplinary ambition. Chapter 1 presents a historical and conceptual foundation, surveying classical and contemporary approaches to emergence and clarifying key distinctions such as ontological vs. epistemological, weak vs. strong, and synchronic vs. diachronic emergence.

Chapter 2 introduces core concepts in complexity science, with particular emphasis on how natural phenomena are interpreted through this lens. It explores how systems are described across different informational levels, preparing the ground for the analytical approach developed in the following chapters.

Chapter 3 presents an original, quantitative definition of emergent properties based on what we term parametric model complexity. This chapter examines how small variations in control parameters can lead to universal system behaviors and offers tools to characterize states and transitions. These elements provide a foundation for formalizing emergence within a dynamic and computational framework.

Chapter 4 explores how this framework can inform scientific reasoning, including theory choice, technological innovation, and the management of anomalies. Well-known historical case studies from the philosophy of science are reinterpreted here—not for their historical content per se, but as illustrative test cases for our central thesis: that increasing parametric complexity can function as an early indicator of epistemic stress.

Finally, Chap. 5 synthesizes the conceptual and methodological contributions of the book. It argues that emergence, complexity, and predictive epistemology together offer a powerful lens through which to understand both the behavior of natural systems and the evolution of scientific knowledge.

Contents

1 The Concept of Emergence 1
1.1 Historical Background 1
1.1.1 Other Ideas 9
1.2 Types of Emergence 10
1.2.1 Ontological Emergence 11
1.2.2 Epistemological Emergence 12
1.2.3 Strong Emergence 13
1.2.4 Weak Emergence 14
1.2.5 Synchronic and Diachronic Emergence 17
References 18

2 Complexity Measures 21
2.1 Choosing a Theory 21
2.2 Complex Systems and the Foundations of Emergence 23
2.2.1 The Landscape of Complexity Science 23
2.3 Complex Systems: Methodologies 24
2.3.1 Non-linear Science 26
2.3.2 Bifurcation Theory 27
2.3.3 Network Theory 31
2.3.4 Game Theory 32
2.3.5 Information Theory 33
2.3.6 Super-Statistic 34
2.3.7 Cellular Automata 35
2.3.8 Agent Base Modelling 36
2.3.9 Data Mining 36
2.4 Complexity Measures 37
2.4.1 Computational Complexity 38
2.4.2 Algorithmic Complexity 39
2.4.3 Effective Complexity 40
2.5 Theories and Models 40
References 41

3 **Complexity and Emergence** ... 45
3.1 States of a System ... 45
3.2 Transitions Between States ... 47
3.3 Complexity Evolution ... 47
3.4 Control Parameters ... 50
3.5 Universality ... 52
3.6 Basin of Attraction ... 53
3.7 Type of Attractors ... 55
3.7.1 Fixed Points ... 56
3.7.2 Attractors of Finite and Infinite Number of Points ... 56
3.8 Variation of the Control Parameter and the Complexity Measure ... 57
3.9 Complexity Change and Emergent Properties ... 58
References ... 59

4 **Evolution of Scientific Theories and Predictive Epistemology** ... 61
4.1 Introduction ... 61
4.2 Around Copernicus ... 61
4.3 Revolutionary Changes ... 62
4.4 The Geocentric Model ... 63
4.5 The Data and the Adjustment of the Model ... 64
4.6 Retrograde Motion ... 65
4.7 Epicycle and Deferent ... 66
4.8 The Change to Simplicity ... 67
4.9 The Paroxysm of the Geocentric Complexity: The Tycho Brahe's Model ... 69
4.10 Kepler's Search for Armory ... 70
4.11 Galileo and the New Worlds ... 72
4.12 Newton's Synthesis ... 73
4.13 Perihelion of Mercury ... 74
4.14 Parametric Complexity and Theory Change ... 76
4.15 Towards a Predictive Epistemology ... 77
References ... 78

5 **Conclusions** ... 81

Chapter 1
The Concept of Emergence

1.1 Historical Background

As it is known, the term "emergence" was used for the first time, associated with a philosophical concept, in "Problems of Life and Mind" by George Henry Lewes (1875). In this foundational work, it is mainly recognized that the whole is often more than the sum of its parts and also that at each level of complexity new, and often surprising, qualities emerge which cannot be attributed, at least directly, to the known properties of the constituents.

To clarify this idea, consider a system that can be described at two distinct levels: a lower level, defined by basic interactions, and a higher level, capturing macroscopic or aggregate behavior. If the system is initially in a state that we designate as m_1 at the lower level and M_1 at the higher level, we can identify emergence in two main ways. First, if M_1 cannot be predicted from m_1, or if additional information must be introduced to account for M_1 despite a comprehensive theory of the lower level, we may say that M_1 is an emergent property. Second, if changes in the system's conditions cause a transition from M_1 to M_2 at the higher level, but our theory of the macroscopic behavior fails to account for this change—again requiring new theoretical resources—we may also consider this shift to reflect an emergent process. These two perspectives capture both vertical emergence (from lower to higher levels) and horizontal emergence (transformations within the higher level that are not internally predictable).

Emergence was well received and adopted by the British school of philosophy in the late nineteenth and early twentieth centuries, particularly in the field of chemistry and biology (Mill, 1843; Alexander, 1920; Morgan, 1923; Broad, 1923, 1925). At those times, vitalism was very present in biology, adhering to the notion that living organisms possessed some form of additional essence (some non-physical element) that animated them. Vitalism came into conflict with orthodox physics, suggesting that organisms were merely highly complex machines, their behavior is

M. Fuentes, *Complexity, Emergence and the Evolution of Scientific Theories: Towards a Predictive Epistemology*, SpringerBriefs in Philosophy,
https://doi.org/10.1007/978-3-031-99994-9_1

ultimately explained in terms of basic physical laws that operate at the molecular level. The emergentists looked for an intermediate position, discarding the vital essences, but denying that all the properties of the living organisms could be completely reduced to, or explained in terms of the (successful) mechanics theory. This would correspond to what is outlined by the vertical emergence in the previous text.

Mid-twentieth century advances in areas such as physics and biology, especially in the elucidation of the fundamental structure of matter, and theories such as quantum physics and the molecular basis of biology, were added to the reductionist approach, explaining many properties of the matter in terms of atomic physics, and many properties of life in terms of molecular mechanisms. This is why, for many, emergence took a second place, becoming an irrelevant anachronism, or perhaps even worse, a vestige of vitalism (Bedau & Humphreys, 2008).

As noted by Bedau (1997) and Bedau and Humphreys (2008), advances in complexity science and computational power have prompted a renewed philosophical interest in the concept of emergence. A pivotal moment in this resurgence was Roger Sperry's A Modified Concept of Consciousness (1969), which laid the groundwork for a new wave of emergentist thinking. As Corning (2002) observes, Sperry was arguably the first to clearly articulate the role of downward causation in complex systems such as the human brain, calling attention to the need for new principles of "cognitive and emergent causation" and "top-down determinism."

Sperry described downward causation as the idea that conscious, subjective properties possess causal power in regulating neural activity: "the conscious subjective properties in our present view are interpreted to have causal potency in regulating the course of brain events; that is, the mental forces or properties exert a regulative control influence in brain physiology" (Sperry, 1976).

This renewed momentum in emergentist theory coincided with the rise of complexity science as a distinct research field, which itself was catalyzed by the development of nonlinear mathematical tools, particularly chaos theory and dynamical systems theory. These innovations enabled scientists to model intricate and unstable dynamics with unprecedented precision, often through high-speed numerical simulations. Such methods, commonly referred to as in silico experiments (i.e., experiments conducted via computational simulation), revolutionized how researchers investigate complex systems.

As a result, the study of emergent properties has gained substantial traction in contemporary literature, especially in the philosophy of mind, complexity science, and computational modeling of unstable systems and artificial life. The interplay between these domains has enriched our understanding of emergence and underscored its central role in explaining phenomena that resist reduction to simpler parts.

The physicist and philosopher of science Mario Bunge has proposed that the emergence is a viable construct, summarizing a pluralistic ontology based on scientific levels. He treats emergence in his paper "Emergence of the mind" (1977), wherein he employs general systems theory to discuss emergence as perceived consciousness, or "mind". His logical progression parts from base definitions of terminology such as property, state, emergence and levels, positing that the central nervous system can be described as a concrete thing, therefore its states and changes of state can be regarded as "state space". He then justifies the concept of an "emergent property", which is that "possessed by a system but not by its components" (ibid), and thus the discussion of consciousness can be viewed as an emergent phenomenon, and a synchronous phenomenon. However, Bunge stresses that emergence is relative, yet simultaneously resultant. Similarly, he clarifies that the brain cannot be compared with social or physical phenomena, as the complex system that is the brain is not considered as being made up of "levels", echoing Descartes: "Again, regarding the brain, they will not be able to imagine anything more plausible than that it is composed of many tiny fibers variously interlaced" (Kenny, 1985). It certainly seems logical to consider the mind from the perspective of complex systems theory, as indeed many other authors have done (Bar-Yam, 1997; Albus & Meystel, 2001).

Bunge set the foundations for the concept of psychoneural emergence, as he discussed monism and dualism as possible bases for the consideration of the phenomenon at great length. Therein he discards dualism in this regard as "unclear, at variance with the general framework of science, and untestable", although, as we will see, dualism can be said to be more widely used nowadays in consciousness science, in as much that the human brain is understood to possess two independent halves of the hippocampus. Bunge also argues that other psychoneural monist perspectives, namely eliminative and reductive materialism, do not pertain to the CNS's specific emergent properties, whereas emergentist materialism, on the other hand, did satisfy Bunge's criteria. The mind, he argued, could be considered within this perspective, thus adding to the emergentist knowledge base.

It bears mentioning – and indeed Bunge himself recognized these limitations – that this theory of mind was produced at an early stage in complex systems theory. Especially, emergent consciousness was still a young concept, and remains young to this day. That being said, consciousness and the CNS's functions continue to be considered emergent, in that they are irreducible, but much of the inner workings of the brain, the CNS and the mind remains to be discovered and understood.

Another interesting discussion on emergence in this wave of literature is chaos theorist James Crutchfield's "Is anything ever new?" (1994), in which he develops Bunge's closing point that the way we perceive knowledge depends also on current knowledge and understanding, which can only be described as concepts that emerge from letters and symbols conditioned, at the same time, by our current knowledge. Thus, he asks "how…scientists – or, for that matter, how adaptive agents evolving in populations – ever "discover" anything new in our worlds" (Crutchfield, 1994). Indeed, other authors have discussed the viability of considering countless phenomena as emergent, novel, irreducible, unexplainable, and/or unpredictable at length; yet we will not enter into that debate here. I will content myself with a brief overview of Crutchfield's calculi of emergence, which he holds can provide the basis for a "statistical complexity" metric to measure a complex system's structure. This, along with his suggested model, a hierarchical machine reconstruction, provides a model for measuring emergent complexity, forming part of what he describes as a "trinity of conceptual tools required for studying the emergence of complexity"; computation, dynamics, and induction. Crutchfield also discusses innovation dynamics, and how they are affected by the system model. Therefore, his work is an indispensable reading when considering contemporary emergentist propositions.

In the philosophy of science, one cannot discuss complexity and emergence without relating these phenomena to the theory of levels, for which one could do a lot worse than to consider Emmeche et al. (1997) thesis "Explaining emergence towards an ontology of levels", another influential treatise of post-modern emergentism. Their proposition commences with the premise that scientific evolutionism precludes the concept of unexplainably, traditionally considered a characteristic of emergence. However, scientific development does not preclude emergence, as Emmeche et al. indicate. Therefore, they propose that "an ontological non-reductionist theory of levels of reality which includes a concept of emergence" is necessary – at least when discussing emergent phenomena in systems to which level theory is applicable, of course. To this end, their paper outlines an eponymous ontology of levels resulting from their analysis and critical discussion of the different types of emergence, principally founded from the conceptual standpoint that evolutionism and materialism represent, similarly to Bunge. Thus they hypothesize the existence of four primary levels: the physical-chemical, the biological, the psychological and the sociological; these levels possess "non-homomorphic inter-level relations". Emmeche et al.'s contribution to existing scientific knowledge on emergence is their proposal of a "devitalization" of emergence, a conceptual perspective that is not without merit. Further conclusions include, but are not limited to, the idea that "Emergence as a single process does not constitute levels but only entities; the emergence of primary levels are a combination of entity constitution and subsequent level constitution, by generation of relations between the entities" (emphasis added).

Next, we come to Goldstein's thesis on emergence as a construct, which is another staple piece of literature in complexity science, defining emergence as a concept arising from self-organization and represented as structures, patterns or properties that are both coherent and new (Goldstein, 1999). The case of the pattern's emergence can be assimilated to the upper-level horizontal change.

In this work, he recognizes that emergence is a phenomenon that exists in many and varying fields of study, which we ourselves will discuss in further detail shortly later in this chapter. He also points out that there are some characteristics that these emergent phenomena have in common, despite differences in field or type of system, such as novelty, coherence, macro level, dynamical and extensiveness, as well as stressing that "there is a great need to ascertain how to channel the process of emergence in constructive directions".

This need is in fact accentuated by "real" examples of emergence, such as the phenomena on complex networks in society; such complex networks present intergroup and intragroup dynamics, on which the network supervenes and depends. Goldstein gives an example of these networks as "the spontaneously arising organizational structures and practices that accompany mergers and acquisitions and the newly shaped strategic alliances that are so rife in our contemporary business world".

All of these adds to the mass of evidence in the literature that emergentism is not only re-emerging but has a clear and tangible use and application – at least for the present moment. And seeing as organizations are, in and of themselves, key producers (or lower-level components) of emergent phenomena, organizational dynamics as a field of academia stands to be greatly affected by the new ground made in the concept of emergentism. This new ground has mainly been made in complexity theory, wherein the study of emergentism has sought to establish a theoretical framework and a methodology that is acceptable for all, thus paving the way for the academic study of phenomena that not only exist but also have a fundamental impact on organizations, companies, institutions, etc.

Furthermore, Goldstein also appraises the issue of unpredictability that is often considered a criterion for emergence, explaining that the "non-analytically solvable nonlinearity of complex systems" means that emergent phenomena will never be completely predictable; therefore, these emergent phenomena will constantly be evolving, never identically resembling their prior representations. He states that it is clear that complexity theory has begun, as the whole concept of predictability itself has been restructured and re-examined in the scientific literature. Thus, the fact that complexity theory makes space and accounts for complex, nonlinear systems as a key concept, provides a basis for unpredictability. This unpredictability ensures the validity of emergence, according to Goldstein, since even an infinite amount of testing and processing cannot predict emergent phenomena satisfactorily; just as quantum mechanics depends on the uncertainty principle, so emergence relies on irreducibility or unpredictability. For example, Newman (2010) sets out an interesting argument for the consideration of strange attractors concept that we will discuss later, suffice to say for the moment that a strange attractor is where the system converges, which has particular characteristics in its dynamics, which they make a place where erratic, unpredictable dynamics are produced as examples of emergent phenomena, and therefore, from a mathematical point of view, the unquestionable unpredictability of emergence is supported, as strange attractors are impossible to predict. Thus, the philosophical argument that unpredictability is a temporary condition or characteristic of emergence, and that eventually, we will possess the necessary functions or systems to predict them, does not hold in all cases. With the production of novel ways of viewing emergence, and the concepts related to it, emergentism has been provided with the necessary foundations for its resurgence as applicable, and most of all useful, philosophical tools.

Goldstein also refers to the traditional definition of complex emergentism, in that it occurs independently of the phenomena that give rise to it – these are understood as occurring on macro and micro levels. As Goldstein notes, attaching a characteristic of causality to the concept of levels in emergence theory is a strategy taken by some authors to argue that the emergent level is, in fact, real. Consider Roger Sperry's discussion of "Macro- versus micro-determinism" (1986), for instance; therein, he sets out the logic for this irreducible causality by pointing out that consciousness or "the mind" emerges from the ordinary workings of the brain's connectors and components, yet the brain can be affected by consciousness itself, which means that consciousness possesses causality over the lower-level systems on which it supervenes. Indeed, Schröder (1998) poses the problem of how emergent phenomena can be considered provisional at all if they have causal power. However,

Sperry's specific brand of downward causation is not completely without issues. The dilemma arises when we consider how exactly how lower-level components can cause a higher-level emergent phenomenon, yet at the same time this higher-level emergent phenomenon possesses causal power over the very components it supervenes on. In any case, there are several perspectives within emergentism – such as Sperry's – that provide foundations for the concept of an emergent macro-level in and of itself.

However, reductionism has challenged the validity of emergentism; Henle predicted that the concept of emergence will be superfluous when a more complete theoretical construct is produced, as this "will be able to predict, deduce and reduce emergent phenomena to micro-level processes" (Henle, 1942). Therefore, the concept of emergence can be viewed as a provisional lens through which we can understand what we currently label emergent phenomena, features, properties or structures while – and only while – our understanding of these complex emergents is incomplete; if we ever were to fully understand all micro-level characteristics, emergentism would be null. This may be true, but to my mind at least, the very nature of science is to seek to complete, sometimes by minuscule step by step process – or at least to complement – knowledge. As scientists or knowledge seekers, we create conceptual tools and theories to this end. Thus Goldstein opines that emergentism has re-emerged in parallel to advanced technology (see Langton, 1986 for more on artificial intelligence), propelled further by new mathematical constructs and research methods, arguing that "contemporary complexity theory is proving capable of prying open the black box of emergence". Temporary or not as the case may be, emergence has regained pertinence and applicability as a tool, especially in complexity theory, where the concept of emergentism has proven extremely useful. Goldstein surmises that in turn: complexity theory is developing the necessary tools, methods, and constructs that render the process of emergence less opaque and, thereby, less prone to the tag of "miraculous." In this way, the re-emergence of emergence can be seen as a real, meaningful process that serves a purpose not only for complexity theory but for the philosophy of science as a whole; indeed, Goldstein puts it nicely (in line with the comments on the evolution of knowledge in science mentioned above): there seems to be no end to the emergence of emergent.

We also believe it is important to consider Jaegwon Kim's seminal "Making sense of emergence"; this 1999 paper is one of the most influential to date, wherein he explores concepts such as (Emmeche et al., 2000) supervenience, irreducibility, and downward causation, albeit from the perspective of philosophy of the mind. In his thesis, he attempts to outline a definition of emergence that may be more widely accepted in the academic community, by taking the concept of a property that can be described as emergent and considering it from the perspective of reductionism. This approach had indeed been taken before, although this time Kim uses a model of reduction that "is significantly different from the classic Nagelian model of reduction". In this sense, Kim's reflections are worthy of consideration and analysis in any account of emergency, contributing new theoretical foundations in a field that, as we have seen in this chapter, has utility and applicability in the philosophies of many sciences and systems of different kinds. However, in my view, it is his contributions to the understanding of the problem of downward causation that represents the greatest advances in emergentist theory. This, given that Kim starts from Sperry's concept of downward causation and clarifies some conceptual doubts that had arisen over the three decades between their times of writing.

The doctrines of emergence and its relation to downward causation as set out by Kim are well worth repeating here and bearing in mind for our discussion:

1. The emergence of complex higher-level entities: Systems with a higher-level of complexity emerge from the coming together of lower-level entities in new structural configurations (the new "relatedness" of these entities).
2. The emergence of higher-level properties: All properties of higher-level entities arising out of the properties and relations that characterize their constituent parts. Some properties of these higher, complex systems are "emergent", and the rest merely "resultant".
3. The unpredictability of emergent properties: Emergent properties are not predictable from exhaustive information concerning their "basal conditions". In contrast, resultant properties are predictable from lower-level information.
4. The unexplainably/irreducibility of emergent properties: Emergent properties, unlike those that are merely resultant, are neither explainable nor reducible in terms of their basal conditions.
5. The causal efficacy of the emergents: Emergent properties have causal powers of their own – novel causal powers irreducible to the causal powers of their basal constituents.

This forms the conceptual basis that consists of a tidy synthesis of emergentist theory since the days of proto-emergence. For an interesting critique on Kim's thesis on downward causation, see "Emergentism, irreducibility, and downward causation" (Stephan, 2003), wherein Achim Stephan discusses different types of emergence (which we will discuss in greater detail below), namely synchronic, diachronic and weak, but this time from the perspective of property reductionism. Thus, he argues that strong diachronic and synchronic theories of emergence are incompatible with this reductionist theory, whereas he believes that this perspective serves only for analyzing weak emergence. Specifically, concerning downward causation, he pays particular attention – as will I – to synchronic emergence, for which I will provide a detailed definition hereunder. Stephan argues,

"For such a theory, a system's property is taken to be emergent if it is irreducible, i.e., if it is not reductively explainable. Furthermore, we have to distinguish two different types of irreducibility with quite different consequences: If, on the one hand, a system's property is irreducible because of the irreducibility of the system's parts' behavior on which the property supervenes, we seem to have a case of "downward causation". This kind of downward causation does not violate the principle of the causal closure of the physical domain. If, on the other hand, a systemic property is irreducible because it is not exhaustively analyzable in terms of its causal role, downward causation is not implied. Rather, it is dubitable how unanalyzable properties might play any causal role at all. Thus, epiphenomenalism seems to be implied."

As we can see, Stephan enunciates the need to differentiate between these two definitions of irreducibility when analyzing emergence, providing a decent grounding on which to base future discussions on downward causation. Indeed, Stephan's understanding of irreducibility will serve us herein for our discussion.

Finally, in the definition and clarification of terms that has formed a significant portion of the wave of literature we have been reviewing in the post-modern re-emergence of emergentist theory, we arrive at de Wolf & Holvoet's, 2004 piece "Emergence versus self-organization: different concepts but promising when combined", which initially sets out to clarify and standardize definitions for both concepts. Their working definition serves as a noteworthy synthesis of emergentist theory over the last several decades, emergence is defined in the following way "A system exhibits emergence when there are coherent emergents at the macro-level that dynamically arise from the interactions between the parts at the micro-level. Such emergents are novel w.r.t. the individual parts of the system". As we can see from the articles mentioned earlier in this chapter, Wolf & Holvoet's definition is very much in-line with the widely accepted understanding of the concept. Next, they proceed to establish a definition for self-organization, which again concisely synthesizes the literature available on the concept; "Self-organization is a dynamical and adaptive process where systems acquire and maintain structure themselves, without external control" (Wolf & Holvoet, 2005).

As the authors point out, both of these phenomena are often shared characteristics in complex systems, and indeed other authors (such as Corning and Kaufmann) have posited that self-organization is, in and of itself, an emergent property. However, de Wolf & Holvoet use emergentist theory to examine self-organization and emergence as independent of each other, such that they can appear independently of each other and also exist in isolation. They point out that the principle similarities between self-organization and emergence are 1. Both are dynamic, arising over time; 2. Both are robust, yet in different ways. For example, emergence is robust because its causal components are flexible so that the macro-level, emergent components will not fail if one micro-level component fails. On the other hand, self-organization is adaptable and can maintain the increased order, and is robust in that sense. Therefore, the only way for us to study emergent macro-level components coherently is to wait for such emergents to self-organize.

Thus, they propose that a combination of the two concepts could be of use when conceptualizing, designing or engineering a coherent behavioral model for multi-agent systems, for these will require the small cause, large effect principle. They argue that there is a focus on non-linear interactivity, which is, in fact, a characteristic of said combination, and which "enables those secondary effects at the macro-level that we call emergent". This nonlinearity is often achieved through positive feedback that amplifies an initial change. The result of the first amplification again triggers positive feedback that amplifies the effect of the change. After a while, several components have 'aligned' themselves with the configuration created by the initial change and the configuration stops growing: the system has 'exhausted' the available resources. This alignment is often the emergent property of the system.

This way, an emergent can self-organize. With this explanation, we can see how the two concepts are indeed distinct and what their use in complexity theory could be.

The authors conclude that, although the two concepts can exist independently, they often occur in parallel in complex dynamical systems; "[i]n such systems, the complexity is huge, which makes it infeasible to impose a structure a priori: the system needs to self-organize. Also, the huge number of individual entities imposes a need for emergence" (emphasis is mine). Reinforcing the need for both concepts, the authors show that emergence and self-organization have already been combined in the literature on large-scale multi-agent systems theory. Therefore, de Wolf & Holvoet add to the wave of literature that draws close links between emergence and complexity, contributing a novel conceptual perspective to strengthen this new field.

1.1.1 Other Ideas

As mentioned at the beginning of this chapter, emergentist theory has been widely applied across the literature. In recent decades, its re-emergence has prompted numerous authors to explore specific types of emergence within their respective fields. One illustrative case is Pier Luigi Luisi's, 2002 article, Emergence in Chemistry: Chemistry as the Embodiment of Emergence, which builds on Emmeche et al., 1997 theory of levels. Luisi presents clear and accessible examples from chemistry to illustrate key conceptual constructs associated with emergence. These range from Sir Francis Crick's reflections on "the soul from a chemical perspective" to classical textbook cases such as the emergent properties of water, the oligomerization of hemoglobin, and the formation of vesicles from surfactant molecules.

As Corning (2002) notes, emergence has been a relevant concept since the rise of molecular biology. However, as Luisi observes, the term itself is rarely invoked within the field. He takes this observation further by examining how central concepts in emergentist theory—such as predictability, explainability, irreducibility, and downward causation—can be analyzed through the lens of chemistry.

Regarding irreducibility, Luisi poses a foundational question: "Can the properties of water be deduced from those of hydrogen and oxygen? Can the aromaticity of benzene be explained in terms of the orbital characteristics of carbon and hydrogen? Can the folding and binding properties of myoglobin be accounted for solely by the properties of its amino acid components?" He argues that in principle, the chemist can answer affirmatively in each case—due largely to the predictive power of quantum chemistry. However, this theoretical deducibility does not negate the emergent status of these properties. Emergence, in this context, is not nullified by deducibility in principle, especially when that deducibility lacks practical tractability.

On the question of predictability, Luisi offers a compelling example:

> In the case of chemistry, the question can be phrased as: in a world where water is not yet known, can water and its properties be predicted on the basis of the properties of hydrogen and oxygen? Likewise: can the existence of myoglobin and its binding properties to oxygen

> be foreseen based on the structure and properties of amino acids? [...] The example of myoglobin helps clarify the point. Given 20 different amino acids (and a chain length of 153 amino acid residues, as in myoglobin), there are approximately 20^{153} (around 10^{200}) possible theoretical chains, and myoglobin is only one of them. In this sense, its primary structure can be predicted in principle, but due to the astronomical size of the theoretical space, it is effectively impossible to identify it among the alternatives. As for its binding properties, these too could be predicted in principle—by calculating the binding behavior of each of the 10^{200} possible chains—but we currently lack the algorithmic means to perform such a task. And it is doubtful whether we will ever achieve it. Nota bene: this is not simply a matter of time or computational power; many scientists acknowledge that the human brain may never be capable of solving this type of problem.

From this perspective, these phenomena can be understood as examples of weak emergence, predictable in principle, but still emergent in practice due to the scale and complexity involved. Most importantly, Luisi also offers a chemical illustration of downward causation:

> Likewise, the linking of many amino acids into a chain to form myoglobin affects the properties of the amino acid components in many respects. To make a peptide bond, two contiguous amino acids must lose one water molecule—a rule that applies broadly to polymerization processes. Furthermore, the folding of the protein affects the physical properties of the amino acid residues; for example, the fluorescence and absorption properties of aromatic residues in the folded chain differ from those of the free amino acids. The fact that these are considered trivial observations by chemists reinforces the view that chemistry is, in fact, the embodiment of emergence—and of downward causality.

As Luisi emphasizes, chemistry offers abundant and concrete examples of downward causation. His thesis stands as a significant contribution to post-modern emergentism by grounding abstract philosophical notions in the empirical realities of chemical systems. His work demonstrates how the core concepts of emergence—irreducibility, predictability, and downward causation—can be fruitfully interpreted within a natural science framework.

Other notable contributors who have extended emergentist theory into specific disciplinary contexts include R. Keith Sawyer, in Social Emergence: Societies as Complex Systems (2005), and Jeffrey Goldstein, in Emergence, Radical Novelty, and the Philosophy of Mathematics (2001). While by no means exhaustive, these examples underscore the conceptual reach and interdisciplinary vitality of emergentist thinking.

1.2 Types of Emergence

Building on the literature discussed earlier in this chapter—as well as the sources examined in the following sections—it becomes evident that emergentist theory is commonly structured around three key classification schemes. These include the distinction between ontological and epistemological emergence; the temporal distinction between synchronic and diachronic emergence; and the gradational distinction between weak and strong emergence. These frameworks have been elaborated by various scholars, including Bedau (1997), Stephan (2003), Gillett (2002), and Chalmers (2006a, 2006b), among others.

1.2.1 Ontological Emergence

Ontological emergentism, also referred to as supervenience emergentism or simply physicalism, is the conceptual framework adopted by British emergentists such as Mill and Broad, as well as by more recent thinkers like Emmeche and collaborators. This school of thought views complexity as organized across multiple levels, with emergent properties arising from interactions between these levels. Crucially, these emergent properties are considered irreducible to the properties of lower levels. However, this irreducibility does not preclude the possibility of downward causation, that is, influences from higher-level phenomena on lower-level processes.

Philosophers such as McLaughlin and Kim interpret this form of emergence as a variant of strong synchronic supervenience. McLaughlin (1997), for instance, offers the following definition: "If P is a property of w, then P is emergent if and only if (1) P supervenes with nomological necessity, but not with logical necessity, on properties the parts of w have taken separately or in other combinations; and (2) some of the supervenience principles linking properties of the parts of w with w's having P are fundamental laws" (p. 39).

This definition highlights key features commonly associated with strong emergence: unpredictability, unexplainability, downward causation, and irreducibility. We will return shortly to explain in more detail why this framework is typically classified as a form of strong emergence.

However, other scholars have raised critical challenges to this account. Notably, Wong (2010), in his article The Secret Lives of Emergents, argues that if emergent properties possess novel downward causal powers—as supervenience emergentism allows—then their co-variation with basal properties is no longer guaranteed. This, he claims, undermines the very notion of supervenience.

Building on this insight, Timothy O'Connor has developed an alternative framework. In contrast to his earlier position (1994), which accepted supervenience as compatible with emergence, his later work (2000a, 2000b) rejects this assumption. He argues that emergence cannot be simultaneously supervenient, causal, and synchronic. Instead, he proposes a dynamic emergentism in which emergent properties are causal but not supervenient, and dynamic rather than synchronic. In this view, both basal and emergent properties contribute to the formation of subsequent properties. As a result, the same physical base might, under different nomological conditions, give rise to different emergent properties—an idea with potential implications for understanding psychological states. However, it remains unclear whether any concrete examples of this mechanism have been empirically demonstrated. For further discussion, see O'Connor and Wong (2005).

Along similar lines, Paul Humphreys has proposed the notion of fusion as an alternative to supervenience. According to Humphreys, "[Emergent properties] result from an essential interaction [i.e., fusion] between their constituent properties—an interaction that is nomologically necessary for the existence of the emergent property." In this account, emergent properties possess novel causal powers, but they do not exist independently of their constituent base properties. As such, their causal autonomy is more limited than that typically granted in supervenience-based accounts. For Humphreys, fusion is not merely a logical or mathematical relation applied to predicates, but rather a real physical process in which new properties arise from the integrated interaction of their components.

1.2.2 Epistemological Emergence

In contrast to ontological emergentism, this school of thought—represented notably by the British philosopher Samuel Alexander—defines emergence in complex systems strictly in epistemological terms, not as a metaphysical concept but as a limitation of human knowledge. Epistemological emergentism is generally divided into two main types. The predictive model refers to properties that cannot be predicted based on pre-emergent properties, even with a deep and exhaustive understanding of the system's basic components and the laws governing them. In contrast, the reductionist model involves systemic properties that give rise to models irreducible to the dynamics of physical theories. These models may support generalizations akin to laws, but they are not laws in a strict sense. It is worth noting that within the domain of epistemological emergence—anticipating a discussion that follows shortly—both diachronic and synchronic phenomena are generally considered possible, and both have been widely addressed in the literature. For instance, some theorists investigate unpredictable macroscopic properties that arise during the dynamic evolution of the world, or the transitions between pre- and post-complexity material states—both examples of diachronic emergence. Others focus on the interplay between micro- and macro-level theories, which are typically regarded as instances of synchronic emergence. This temporal dimension of emergence will be examined in more detail shortly.

We should also mention the irreducibility perspective on emergence, which stems from the work of Ernest Nagel (1961, pp. 366–374). Paul Teller builds on this line of reasoning in A Contemporary Look at Emergence (1992), offering the following definition: "A property is emergent if and only if it is not explicitly definable in terms of the non-relational properties of any of the object's parts" (pp. 140–141). Teller further emphasizes that his definition encompasses both relational and non-relational properties. This irreducibility criterion has already appeared earlier in this chapter, as it is often considered a central condition for identifying emergent phenomena.

1.2.3 Strong Emergence

In strong emergence, the issue that is most relevant, from a metaphysical perspective, is that which considers the question of exactly what criteria must be fulfilled for an emergent property to qualify as truly, genuinely distinct. In other words, the question is can we draw a scientifically exact distinction –from a metaphysical perspective– between components that are in fact "nothing over and above" their constituent components, and components that can be considered as "something over and above" their components. The concept of strong emergence that prevails in philosophy aims to represent this ontological distinction and helped form an influential concept in British emergentism. Modern authors such as Bedau, as mentioned in the previous section, have added the concept of weak emergence in contrast, and we will discuss this further in the next section, which is devoted to weak emergence.

Although British Emergentism authors such as C. D. Broad defined strongly emergent properties as ones that are "not deducible from knowledge of the lower-level domain from which [they] emerged", what actually distinguishes the concept of strongly emergent (novel) properties from that generally taken in epistemological emergentism (as discussed above, weak emergence tends to prevail here), is in fact the result of an ontological distinction. Therefore the criteria for a higher-level component to be considered novel must first be laid out for the proper classification of a property to be defined as strongly emergent.

Despite being considered equal to the sum of the related properties of its parts, a component may be considered novel – even according to reductive physicalists. For example, the sum of the weights of its parts is as much as the weight of the whole, but the parts do not share the weight of the whole. From this perspective, the criteria in strong emergence would not consider the weight – the novel resultant property – as strongly emergent. Strong emergence also provides a conceptual distinction between many resultant properties and truly emergent properties, because numerical novelty, such as the first-time a given shape or mass occurs in nature, is not sufficient for the property to be considered strongly emergent.

For this reason, the concept of strong emergence generally holds causal novelty as one of the necessary criteria –as well as irreducibility. That is to say that the fundamental principles of strong emergence are that a higher level component is irreducible to the lower-level components on which it supervenes and that it has causal power over them; in other words, they possess downward causal powers, as opposed to resultant causal powers). Therefore, the British emergentists and other ontological emergentism authors have generally studied novel emergent laws as well as properties, subscribing to the theory of levels and downward causal powers. Strongly emergent properties were succinctly defined by O'Connor (1994) – despite the fact that he subsequently repudiated this stance in 2000 like properties that supervene on a complex object's lower-level parts, but are not shared with them, that are distinct and novel, and that possess downward causal powers over the lower-level components' behavior.

However, as with many of these concepts, there is no clear consensus in the literature on strong emergence, and some theorists do not require downward causation for strong emergence; see Chalmers' "Strong and weak emergence" (2006a, 2006b) for an example of this take on the issue. His definition of strong emergence also bears mentioning here: "a high-level phenomenon is strongly emergent when it is systematically determined by low-level facts but truths concerning that phenomenon, which is in principle not deducible from truths in the lower-level domain". From his perspective, emergent phenomena are considered supervenient on lower-level components or facts in a nomological sense, but not a logical sense; thus, novel fundamental laws are needed to understand how the characteristics of components in different levels are related to each other.

Again, a slightly different definition of the concept of strong emergence is proposed by Tim Crane (2001, 2010), who sets out his theoretical framework based on the distinction between two types of reduction: (1) ontological reduction, which links higher-level components or entities with lower-level components; and (2) explanatory reduction: a link between theories to understand the higher-level phenomena in terms of the lower-level phenomena. His example is that "one theory, T2, is explanatorily reduced to another, T1, when theory T1 sheds light on the phenomena treated in T2; that is, shows from within theory T1 why T2 is true". Crane distinguishes between non-reductive physicalism and strong emergentism from the perspective of this theoretical framework; the former can be seen to encompass the concept of explanatory reduction whilst denying ontological reduction, whereas the latter – strong emergence – denies both types of reduction, therefore leaving an explanatory gap. His principle posits that if supervenience exists, as well as a priori irreducibility and downward causation, then there is dependence without explanatory reduction and, consequently, we can say that strong emergence occurs.

1.2.4 Weak Emergence

As mentioned earlier, strong emergence was added to the conceptual melting pot relatively recently by modern authors, yet weak emergence existed before this; we can trace its roots back to British Emergentism. Weak emergence is the kind that is most common in the early twenty-first century, with notions such as complexity, functional organization, self-organization and non-linearity coming to the fore – as seen above – mostly in the fields of cognitive science, complex systems theory and, in broader terms, various scientific discussions about emergence. Central to this understanding of a weakly emergent property is the requisite property of a system as a whole, that none of its smaller parts share, which is unpredictable or unexpected given the properties and the laws governing the lower-level, more fundamental, system or components from which it emerged. Due to this pre-requisite of unpredictability or unexpectedness, weak emergence is not so much a metaphysical concept, but rather it can be considered epistemological. In the literature on weak emergence, the examples of phenomena that are often provided include traffic jams, the flocking patterns of birds, phase transitions, systemic properties of connectionist networks, emergent patterns in cellular automata, termite organisation, et cetera.

The principal contrast with strong emergence is that weak emergence is compatible with reduction due to the fact that within its theoretical framework, it allows for a component to be simultaneously reducible and unpredictable. For example, minute initial conditions can mean that processes with strict deterministic laws may still be unpredictable due to unforeseeable consequences. Chalmers uses the example of deducibility that can be seen with cellular automata, wherein the lower-level components and rules can be used to deduce higher-level phenomena or properties, yet these remain unexpected, to set the foundations for weak emergence to be considered compatible with the deductibility of an emergent phenomenon.

Another example of weak emergence is Mario Bunge's "rational" emergentism (and we revisit "Emergence of the mind", 1977), as he sets out a theoretical framework wherein emergent properties follow this definition of systemic properties as a whole possessing characteristic that none of the system's parts share; yet these are reducible to the system's parts and the way they are organised. He justifies the concept of an "emergent property", which is that "possessed by a system but not by its components" (ibid), and thus the discussion of consciousness can be viewed as an emergent phenomenon, and a synchronous phenomenon at that. It is worth noting, however, that according to this theory Bunge considers emergence to be relative, yet simultaneously resultant. He claims that this form of emergentism differs from the reductionist mechanism since it considers the fact that systemic properties may be novel within its conceptual framework. Furthermore, Bunge argues for the existence of a reductive explanation for this novelty. This he dubbed as "rational" emergence.

William Wimsatt (1997) also defends an account according to which emergence is compatible with reduction. Wimsatt defines emergence negatively as the failure of aggregativity; aggregativity is the state in which "the whole is nothing more than the sum of its parts" in which, that is, systemic properties are the result of the component parts of a system rather than their organization. Contrasting emergence to aggregativity, Wimsatt defines a systemic property as emergent relative to the properties of the parts of a system if the property is dependent on their mode of organization (and is also context-sensitive) rather than solely on the system's composition. He argues that, in fact, it is aggregativity which is very rare in nature, while emergence is a common phenomenon (even if in different degrees).

Robert Batterman (2001), who focuses on emergence in physics, also believes that emergent phenomena are common in our everyday experience of the physical world. According to Batterman, what is at the heart of the question of emergence is not downward causation or the distinctness of emergent properties, but rather intertheoretic reduction and, specifically, the limits of the explanatory power of reducing theories. Thus, a property is emergent, according to this view, if it is a property of a complex system at limit values that cannot be derived from lower level, more fundamental theories. As examples of emergent phenomena Batterman cites phase transitions and transitions of magnetic materials from ferromagnetic states to paramagnetic states, phenomena in which novel behavior is exhibited that cannot be reductively explained by the more fundamental theories of statistical mechanics.

However, Batterman wants to distinguish explanation from reduction and so claims that though emergent phenomena are irreducible they are not unexplainable per se because they can have non-reductive explanations.

More recently Mark Bedau (1997, 2008) has argued that the characteristic of weak emergence is that, though macro-phenomena of complex systems are in principle ontologically and causally reducible to micro-phenomena, their reductive explanation is intractably complex, save by derivation through simulation of the system's microdynamics and external conditions. In other words, though macro-phenomena are explainable in principle in terms of micro-phenomena, these explanations are incompressible, in the sense that they can only be had by "crawling the micro-causal web"—by aggregating and iterating all local micro-interactions over time. Bedau argues that this is the only kind of real emergence and champions what he calls the "radical view" of emergence according to which emergence is a common phenomenon that applies to all novel macro-properties of systems. (He contrasts this to what he calls the "sparse view" which he characterizes as the view that emergence is a rare phenomenon found only in "exotic" phenomena such as consciousness that are beyond the scope of normal science.) However, though this is a weak kind of emergence in that it denies any strong form of downward causation and it involves reducibility of the macro to the micro (even if only in principle), Bedau denies that weak emergence is merely epistemological, or merely "in the mind" since explanations of weak emergent phenomena are incompressible because they reflect the incompressible nature of the micro-causal structure of reality which is an objective feature of complex systems.

Andy Clark (1997, 2001) also holds a weak emergentist view according to which emergent phenomena need not be restricted to unpredictable or unexplainable phenomena but are, instead, systemic phenomena of complex dynamical systems that are the products of collective activity. Clark distinguishes four kinds of emergence. First, emergence as collective self-organization (a system becomes more organized due solely to the collective effects of the local interaction of its parts, such as flocking patterns of birds, or due to the collective effects of its parts and the environment, such as termite nest building). Second, emergence as unprogrammed functionality, that is, emergent behaviour that arises from repeated interaction of an agent with the environment, such as wall-following behaviour in "veer and bounce" robots (Clark, 1997). Third, emergence as interactive complexity in which effects, patterns or capacities of a system emerge resulting from complex, cyclic interaction of its components. For example, Bénard and Couette convection cells that result from a repetitive cycle of movement caused by differences in density within a fluid body in which the colder fluid forces the warmer fluid to rise until the latter loses enough heat to descend and cause the former fluid to rise again, and so on. And fourth, emergence as uncompressible unfolding (phenomena that cannot be predicted without simulation). All of these formulations of emergence are compatible with reducibility or in principle predictability and are thus forms of weak emergence. For Clark, emergence picks out the "distinctive way" in which factors conspire to bring about a property, event or pattern and it is "linked to the notion of what variables figure in a good explanation of the behavior of a system." Thus, Clark's notion of emergence in complex systems theory is explanatory in that it focuses on explanations in terms of collective variables, that is, variables that focus on higher-level features of complex dynamical systems that do not track properties of the components of the system but, instead, reflect the result of the interaction of multiple agents or their interaction with their environment.

Proponents of weak emergence do not support the strong notion of downward causation that is found in strong emergentist views but, instead, favor one in which higher-level causal powers of a whole can be explained by rules of interaction of its parts, such as feedback loops. Though this kind of view of emergence is predominant in the sciences, it is not exclusive to them. A form of weak emergence within philosophy that denies strong downward causation can be found in John Searle (1992). Searle allows for the existence of "causally emergent system features" such as liquidity, transparency and consciousness that are systemic features of a system that cannot be deduced or predicted from knowledge of causal interactions of lower levels. However, according to Searle, whatever causal effects such features exhibit can be explained by the causal relations of the systems parts, for example, in the case of consciousness, by the behavior and interaction of neurons.

If we make use, for more precision, of the distinction between ontological and explanatory reduction we can see that if we understand strongly emergent phenomena as both ontologically and explanatorily irreducible, as Crane (2010) does, then they are also weakly emergent. However, if strongly emergent phenomena are only ontologically irreducible they may still be, in principle, predictable. For example, even if you deny the identity of heat with mean kinetic energy (perhaps because of multiple realizability) a Laplacean demon could still predict a gas' heat from the mean kinetic energy of its molecules with the use of "bridge laws" that link the two vocabularies. These bridge laws can be considered to be part of what Crane calls an explanatory reduction. So in such cases, strong emergence does not entail weak emergence. Also it should be noted that weak emergence does not entail strong emergence. A phenomenon can be unpredictable yet also ontologically reducible: perhaps for instance, because systemic properties are subject to indeterministic laws. So a case of weak emergence need not necessarily be a case of strong emergence.

Another distinction that is made concerning how novelty is understood is the distinction between synchronic and diachronic novelty. The former is novelty exhibited in the properties of a system vis-à-vis the properties of its constituent parts at a particular time; the latter is temporal novelty in the sense that a property or state is novel if it is instantiated for the first time. This leads to distinction between synchronic and diachronic emergence.

1.2.5 Synchronic and Diachronic Emergence

In synchronic emergence, articulated by C. D. Broad and predominant in the philosophy of mind, the higher-level, emergent phenomena are simultaneously present with the lower-level phenomena from which they emerge. Usually, this form of emergence is stated in terms of supervenience of mental phenomena on subvenient/subjacent neural structures, and so mental states or properties co-exist with states or properties at the neural level. Strong ontological emergence is thus usually understood to be synchronic, "vertical", emergence. In contrast, diachronic emergence is "horizontal" emergence evolved through time in which the structure from which the novel property emerges exists before the emergent. This is typical of the weakly emergent states appealed to in discussions of complex systems, evolution, cosmology, artificial life, and so forth. It can be found in Searle (1992) since he views the relation of the emergent to its base as causal thus, at least in non-synchronic accounts of causation, excluding synchronic emergence.

Because diachronic emergence is emergence over time, novelty is understood in terms of the unpredictability of states or properties of a system from past states of that system. And because weak emergence is typically defined in terms of unpredictability it is also usually identified with cases of diachronic emergence. In contrast, in synchronic emergence, which refers to the state of a system at a particular time, novelty revolves around the idea of irreducibility and thus synchronic emergence is usually identified with strong emergence. However, there are formulations of non-supervenience-based strong emergence that are causal and diachronic, such as O'Connor and Wong's (2005). Note that synchronic emergence could be the result of diachronic emergence but is not entailed by it since, presumably, if God were to create the world exactly as it is in this moment, synchronically emergent phenomena would exist without them being diachronically emergent.

References

Albus, J. S., & Meystel, A. (2001). *Engineering of mind: An introduction to the science of intelligent systems*. Wiley.

Alexander, S. (1920). *Space, time, and deity* (Vol. 2). Macmillan.

Bar-Yam, Y. (1997). *Dynamics of complex systems*. Addison-Wesley.

Batterman, R. W. (2001). *The devil in the details: Asymptotic reasoning in explanation, reduction, and emergence*. Oxford University Press.

Bedau, M. A. (1997). Weak emergence. *Noûs, 31*(s11), 375–399.

Bedau, M. A. (2008). Is weak emergence just in the mind? *Minds and Machines, 18*(4), 443–459.

Bedau, M. A., & Humphreys, P. E. (2008). *Emergence: Contemporary readings in philosophy and science*. MIT Press.

Broad, C. D. (1923). *Scientific thought*. Humanities Press.

Broad, C. D. (1925). *The mind and its place in nature*. Routledge.

Bunge, M. (1977). Emergence and the mind. *Neuroscience, 2*(4), 501–509.

Chalmers, D. J. (2006a). *The re-emergence of emergence. Chapter strong and weak emergence*. Oxford University Press.

Chalmers, D. J. (2006b). *Strong and weak emergence. The reemergence of emergence* (pp. 244–256). Oxford University Press.

Clark, A. (1997). *Being there*. MIT Press.

Clark, A. (2001). *Mindware*. MIT Press.

Crane, T. (2001). The significance of emergence. In C. Gillett & B. M. Loewer (Eds.), *Physicalism and its discontents* (pp. 207–224). Cambridge University Press.

Crane, T. (2010). Cosmic hermeneutics vs. emergence: The challenge of the explanatory gap. In *Emergence in mind* (pp. 22–34).
Corning, P. A. (2002). The re-emergence of emergence: A venerable concept in search of a theory. *Complexity, 7*(6), 18–30.
Crutchfield, J. P. (1994). Is anything ever new? Considering emergence. SFI working paper: 1994-03-011.
De Wolf, T., & Holvoet, T. (2004). Emergence versus self-organisation: Different concepts but promising when combined. In *International workshop on engineering self-Organising applications* (pp. 1–15). Springer.
Emmeche, C., Køppe, S., & Stjernfelt, F. (1997). Explaining emergence: Towards an ontology of levels. *Journal for General Philosophy of Science, 28*(1), 83–117.
Emmeche, C., Finnemann, N. O., & Christiansen, P. V. (2000). In P. B. Andersen (Ed.), *Downward causation: Minds, bodies and matter*. Aarhus University Press.
Gillett, C. (2002). The varieties of emergence: Their purposes, obligations and importance. *Grazer Philosophische Studien, 65*, 95–121.
Goldstein, J. (1999). Emergence as a construct: History and issues. *Emergence, 1*(1), 49–72.
Henle, P. (1942). The status of emergence. *The Journal of Philosophy, 39*(18), 486–493.
Kenny, A. (1985). *The philosophical writings of Descartes* (Vol. 1). Cambridge University Press.
Langton, C. G. (1986). Studying artificial life with cellular automata. *Physica D: Nonlinear Phenomena, 22*(1–3), 120–149.
Lewes, G. H. (1875). Problems of life and mind. Forgotten Books, 2016.
Luisi, P. L. (2002). Emergence in chemistry: Chemistry as the embodiment of emergence. *Foundations of Chemistry, 4*(3), 183–200.
McLaughlin, B. (1997). Emergence and supervenience. *Intellectica, 25*(1), 25–43. Emergence and Supervenience.
Mill, J. S. (1843). *A system of logic, ratiocinative and inductive*. Longan and Green.
Morgan, C. L. (1923). *Emergent evolution*. Henry Holt.
Nagel, E. (1961). The structure of science New York. *Harcourt, Brace and Worldfip, 74*, 18–31.
Newman, M. (2010). *Networks: An introduction*. Oxford University Press, Oxford.
O'Connor, T. (1994). Emergent properties. *American Philosophical Quarterly, 31*(2), 91–104.
O'Connor, T. (2000a). Persons and causes: The metaphysics of free will. Oxford University Press on Demand.
O'Connor, T. (2000b). Causality, mind, and free will. *Noûs, 34*(s14), 105–117.
O'Connor, T., & Wong, H. Y. (2005). The metaphysics of emergence. *Noûs, 39*(4), 658 678.
Schröder, J. (1998). Emergence: Non deducibility or downwards causation? *The Philosophical Quarterly, 48*(193), 433–452.
Searle, J. R. (1992). *The rediscovery of the mind*. MIT Press.
Sperry, R. W. (1969). A modified concept of consciousness. *Psychological Review, 76*(6), 532–536.
Sperry, R. W. (1976). Mental phenomena as causal determinants in brain function. In *Consciousness and the brain* (pp. 163–177). Springer US.
Sperry, R. W. (1986). Macro-versus micro-determinism. *Philosophy of Science, 53*(2), 265–270.
Stephan, A. (2003). Emergentism, irreducibility, and downward causation. *Grazer Philosophische Studien, 65*(1), 77–93.
Teller, P. (1992). A contemporary look at emergence. Emergence or reduction. (pp. 139–153). DE GRUYTER.
Wimsatt, W. C. (1997). Aggregativity: Reductive heuristics for finding emergence. *Philosophy of Science, 64*, S372–S384.
Wolf, T., & Holvoet, T. (2005). Towards a methodology for engineering self-organising emergent systems. *Frontiers in Artificial Intelligence and Applications, 135*, 18.
Wong, H. Y. (2010). The secret lives of emergents. *Emergence in Science and Philosophy*, 7–24.

Chapter 2
Complexity Measures

2.1 Choosing a Theory

Before the twentieth century, there was broad consensus that a large number of degrees of freedom in a physical system was a necessary condition for unpredictability in mechanical, chemical, or biological systems. One of the most influential expressions of this view was formulated by Pierre-Simon Laplace, who wrote: "We may regard the present state of the universe as the effect of its past and the cause of its future. An intellect which at a certain moment would know all forces that set nature in motion, and all positions of all items of which nature is composed, if this intellect were also vast enough to submit these data to analysis, it would embrace in a single formula the movements of the greatest bodies of the universe and those of the tiniest atom; for such an intellect, nothing would be uncertain and the future, just like the past, would be present before its eyes." (Laplace, 1825).

This passage captures a strictly deterministic worldview—one in which surprise or the emergence of new properties is impossible, since Laplace's demon could, in principle, calculate all future states of the universe with perfect precision.

Earlier iterations of this notion appear in the work of Gottfried Leibniz, who wrote:

> Everything proceeds mathematically... If someone could have a sufficient insight into the inner parts of things, and in addition had remembrance and intelligence enough to consider all the circumstances and take them into account, he would be a prophet and see the future in the present as in a mirror.

Likewise, Cicero (106–43 BCE) offers a similar reflection:

> Besides, since everything happens by fate... if there could be any mortal who could observe with his mind the interconnection of all causes, nothing indeed would escape him. For he who knows the causes of things that are to be, necessarily knows all the things that are going to be. But since no one but God could do this, what is left for man is that he should be aware of future things in advance by certain signs... For the things which are going to be

M. Fuentes, *Complexity, Emergence and the Evolution of Scientific Theories: Towards a Predictive Epistemology*, SpringerBriefs in Philosophy,
https://doi.org/10.1007/978-3-031-99994-9_2

> do not come into existence suddenly, but the passage of time is like the unwinding of a rope, producing nothing new but unfolding what was there at first, (On Divination).

Today, however, we know that even deterministic systems with very few degrees of freedom can exhibit highly complex and unpredictable behavior, especially when chaos (either classical or quantum) is present. As the saying attributed to Niels Bohr goes, "Prediction is very difficult, especially about the future." Similarly, Feynman famously remarked: "The next great awakening of human intellect may well produce a method of understanding the qualitative content of equations… Today we cannot see whether Schrödinger's equation contains frogs, musical composers, or morality – or whether it does not," (Feynman et al., 1964).

Bohr's quote alludes to the uncertainty principle in quantum mechanics, where, even with the most accurate theories, future outcomes are only probabilistically determined. This imposes an inherent limitation on predictive power.

Feynman's comment, meanwhile, points to a different but related issue: the challenge of extracting meaningful, interpretable content from highly abstract equations. Even in fully deterministic theories, such as those used to describe fluid dynamics, the sheer complexity of the equations may obscure the emergence of certain qualitative features. This highlights a fundamental epistemological concern: the relationship between formal mathematical representation and human comprehension.

Feynman also draws attention to the fact that many theories, especially in physics, describe not only the actual behavior of a system, but also a vast landscape of potential or hidden states. Exploring these states requires immense computational power, and doing so effectively is a central challenge in the epistemology of complex systems. The analysis of such issues, grounded in quantitative tools from information theory and complexity science, lies beyond the scope of this chapter, but will be addressed in future work.

In light of these reflections, a key question arises: To what extent does a given theory accurately describe the phenomenon it purports to explain? In other words, we must consider the coarse-graining level at which the theory remains valid and predictive. A theory can be said to "work" if it successfully captures the regularities of interest at a given level of abstraction. For example, classical Newtonian mechanics explains planetary orbits with remarkable accuracy, but fails to account for subtle effects such as the anomalous precession of Mercury's perihelion, an effect that only general relativity can explain.

Of course, explanatory adequacy is not the only criterion a theory must satisfy. From a theoretical perspective, explanations should also be concise. In this book, we will employ a measure central to modern epistemology and information theory: Kolmogorov complexity, denoted K, which refers to the length of the shortest computer program that can reproduce a given output. A good scientific explanation should have low K, that is, it should be as simple and compact as possible. Moreover, it should not predict every possible outcome indiscriminately; rather, it should capture the salient regularities of the observed phenomenon. At the same time, the entropy H (a measure of uncertainty or ignorance about the system) should also be low.

These principles, simplicity, specificity, and minimal uncertainty, provide foundational support for the epistemological framework developed in this book. They will be examined in more detail in subsequent sections.

To properly engage with these ideas and the complexity measures they involve, it is essential to understand the methodological landscape of complexity science. This field offers a rich toolkit for examining the emergence of novel properties, evaluating the explanatory power of different theoretical models, and clarifying the relationship between macroscopic phenomena and their microscopic underpinnings. In the rest of this chapter, we provide a brief overview of these methodologies, as a foundation for exploring the links between emergence, complexity, and the epistemology of scientific explanation.

2.2 Complex Systems and the Foundations of Emergence

A widely acknowledged connection exists between complex systems and the phenomenon of emergence, a relationship deeply embedded in the discourse of contemporary complexity science (Bedau & Humphreys, 2008). This connection becomes clear when we conceptualize a complex system as an ensemble of interacting elements governed by simple local rules. These interactions, often non-linear and sensitive to environmental feedback, can give rise to macroscopic behaviors that are profoundly unexpected from the perspective of the system's individual components. Such behaviors are typically referred to as emergent properties or emergent phenomena.

As emphasized by thinkers like Lewes (as reformulated by Nicolis & Nicolis, 2012), non-linearity and context-sensitive interactions are necessary preconditions for emergence. The collective properties that arise in these systems are not merely the sum of their parts but constitute novel, often surprising features that cannot be easily deduced from lower-level dynamics alone. In this sense, complexity and emergence are intimately linked, not only conceptually, but also methodologically.

In the sections that follow, we will explore the defining characteristics of complex systems and introduce key quantitative tools for measuring complexity. This foundation will allow us to formalize the intuitive connection between complexity and emergence, bridging philosophical insight with computational rigor.

2.2.1 The Landscape of Complexity Science

Complexity science has emerged as a major force across the scientific landscape, not as a traditional discipline, but as an inherently interdisciplinary or even transdisciplinary field. Its reach spans a remarkable diversity of domains, from the physics of plasmas to the evolution of human languages, from the self-organization of cellular structures to the dynamics of economic and ecological systems. What unifies these domains is not subject matter, but methodology: the investigation of collective phenomena that arise from decentralized interactions among components.

Unlike earlier scientific revolutions that focused on the extremes of scale, the very small (as exemplified by quantum mechanics following Planck's 1900 discoveries) and the very large (as radically reconceived by Einstein's theory of relativity in 1905), complexity science targets the meso-level of organization. It is at this intermediate scale, where many-body interactions unfold in nonlinear and adaptive ways, that some of the most puzzling and generative patterns in nature emerge.

Indeed, complexity science marks a shift in the scientific frontier: away from reductionist descriptions of isolated components, toward integrative models that seek to understand how relational structures and feedback loops shape system-wide behavior. Increasingly, researchers are attempting to describe the collective dynamics of systems in physics, chemistry, biology, and social science using quantitative methods capable of capturing emergent behavior.

This shift has profound epistemological implications. It challenges classical assumptions about prediction, control, and explanation, and invites a new set of tools—drawn from information theory, dynamical systems, and network science—to make sense of the richness and unpredictability of complex systems. In doing so, it provides a rigorous framework for understanding the nature and origins of emergent phenomena.

2.3 Complex Systems: Methodologies

Although there is no precise and, therefore, unique definition of complex systems, most researchers agree on some of the essential properties that a system must possess to be called a complex (Boccara, 2010). A complex system:

- consists of a large number of interacting agents, usually through simple rules;
- shows emergent characteristics: that is to say, a collective behavior difficult to predict (it does not result from the existence of a central controller), this type of behavior is usually called self-organized (Miller & Page, 2007).

An interesting and relatively complete discussion of these characteristics (presented by any complex system) and the mathematical models that could be used as an approximation for them can be found in Nicolis and Nicolis (2012). The basic and important idea presented there is that non-linear behavior is a *necessary condition* for a complex behavior, and a characteristic of this is the multiplicity of different states that the system can explore, some of them using the same set of parameters.

The diversity and specificity of these properties and characteristics explain the great variety of concepts that are linked to complexity: self-organization, self-organized criticality, self-similarity (typical of fractal objects), etc. The challenges to understand these properties continue, and reduce the hope of any completely unified vision of the domain of complexity, although undoubtedly efforts in this direction continue.

Before analyzing in detail some methodologies of complex systems, we will cite some simple examples for to have a more near approximation to it. As we mentioned, according to Nicolis and Nicolis (2012), in terms of the mathematical modeling of a physical system, the non-linearity in the evolution equations is a necessary condition for complex behavior. This leads, in some cases, to the fact that the system is in bifurcation zones (Guckenheimer & Holmes, 2013), which will ultimately result in novel features and behavior resulting from the jump, via the aforementioned bifurcation, to new solutions (stationary or not). In effect, the theory of bifurcations (after the seminal work of H. Poincaré in 1885), describes the structural changes in the mathematical solutions of dynamic systems when making small changes in the conditions of the equations. This is the basic scenario according to which Nicolis and Nicolis present the possibility where a system is complex or not.

However, John H. Holland (1998), understands by complex system a network of interacting elements, those that evolve in parallel and react constantly according to their state and other elements of the network. This decentralized evolution leads, in certain cases, to surprising collective characteristics (emergence), which, a priori, are difficult or impossible to predict, taking into account only the evolution of individual agents.

While the first definition of a complex system focuses on the dynamic properties of the system in the vicinity of bifurcation points, the second definition refers to the interaction of many elements, evolving individually and collectively (Bonabeau, 2002).

As can be seen, there are difficulties when describing a given system as complex, since under different perspectives a system is complex or not. This difficulty has not been overcome, although clearly complexity sciences continue to produce scientifically acceptable results.

We will note that, at a minimum, there are two types of problems at play. On the one hand, there are the empirical ones that scientists face, such as the prediction of the behavior of complex systems, the interaction between their elements (the coarse-graining) and the causal correlation that this interaction maintains with the behavior of the system as a whole, and the other problems associated with theoretical research when studying empirical systems already recognized as complex by one or the other of the definitions.

But, unlike this type of vagueness, well identified by the scientists of the discipline or disciplines interested in such systems, there is another kind of question that requires attention. These are the conceptual problems (Laudan, 1986) that the discipline faces when scientists study empirical-border problems, for which new tools of explanation, prediction, intervention or application are being elaborated. The conceptual problems are, for example: what are the characteristics that will be taken as defining to decide that a system is complex; and on the other hand, what are, if any, the necessary and sufficient conditions to decide to include a certain case as a complex system; what are, if any, the different types of complex systems and what are the characteristics that define each of these types and, finally, what are the groupings of types and subtypes under certain criteria that provides a taxonomy that sheds light on the diversity of cases.

All these conceptual problems are recently breaking into the physics scene (among others) and it is necessary to continue making efforts to approach a classification that is at the same time consistent, exhaustive, complete and that, in turn, allows us to understand in a more precise and profound way the alluded diversity.

Given that emergent properties are strongly related to the complexity of the system, this concept is heir to the difficulties in its definition and, in turn, the notion of emergence seems to present its nontrivial difficulties for its empirical and conceptual elucidation, as we advance in the preceding chapter.

In the following sections of this chapter, I present some of the methods used in the science of complexity. It will be seen that the techniques are varied and involve different methodologies that study different characteristics and scales of the system. It is also noteworthy that there are important differences, epistemologically speaking, since some methods are completely analytical, while some are numerical, and there are also some that share both types of methodology.

It is worth noting that some of the methods discussed here have a long-standing tradition in physics and mathematics, while others—such as contemporary network theory, despite its historical roots in Euler's graph theory (1736)—are relatively recent developments. We have chosen to set aside discussions related to artificial intelligence for the purposes of this text, focusing instead on theories that can be formulated through mathematical equations and algorithmic models. Nevertheless, the intersection between AI and complexity remains a fascinating area of inquiry, one we intend to address in future work. What truly distinguishes the complex systems approach is its ability to breathe new life into these traditional methodologies, offering a novel and quantitatively rich lens through which to explore natural ways grounded in a deeply interdisciplinary perspective.

2.3.1 Non-linear Science

Scholars from a few decades ago had an established idea: for a given system (or phenomena) subject to a set of conditions -say temperature, pressure, etc., for physical systems, or population size and mean education degree in case of human societies- slight changes on these conditions produce also small, or similarly unimportant, changes in the final state of the system. And also, when studying the superposition of effects on the system, the expected result of two or more actions on the system will be the simple superposition of each effect taking into account each action separately.

The aforementioned properties are the ones taking place in a linear world.

Fortunately, linear systems are in general very rare, even though some important dynamical equations are linear (e.g. the Schrodinger equation, in quantum physics). Many body systems, such as complex and human societies, are highly non-linear. Is means that in this type of systems abrupt transitions can be observed, i.e. the state of the system changes dramatically upon small perturbations. For instance, it can collapse, go extinct or thrive. In some cases, multiple possible stable solutions can arise; and also, unpredictability in both, space and time, which is deterministic systems, is known as classical chaos (Strogatz, 2014).

Another issue that arises, sometimes related to thermodynamics, is scaling. An important attribute of power laws. That is, given a relation of a mathematical function, when scaling the argument by a constant factor, it causes only a proportionate change of the function itself. It is easy to see that when logarithms are taken on the function and the argument, this behavior is what produces a linear relationship. The importance of this simple relation is that the equivalence of power laws with a particular scaling exponent sometimes can have a deeper origin in the dynamical processes that generate this behavior at a microscopical level. The critical exponent, as it is usually referred to in physics, is associated with phase transitions in thermodynamical systems.

In complex systems science, it is very usual to find this particular characteristic. Scaling laws appear in: biological systems (for example the relation between metabolic rate and the size of an organism) (West et al., 1997); fractals; social interactions; cities (an example can be the total road length as a function of population size) etc.

Non-linear science is a cornerstone in complexity studies. Note that the variety, and unpredictability, of the solutions in non-linear systems sometimes is referred as emergent behavior (Bedau & Humphreys, 2008), something that is usual in social and biological systems (Scheffer et al., 2009).

2.3.2 *Bifurcation Theory*

As I mentioned before, non-linear behavior is the usual type of dynamics observed in the many-body systems (social, for example). In these systems, the final stable solutions (equilibrium points or final states) can change drastically when some of the parameters (usually called in the literature "control parameters"), that are relevant to the evolution of the system, reach a particular value. To understand this property, common to almost all non-linear systems, we will discuss some classic bifurcations (Hale & Koçak, 2012; Guckenheimer & Holmes, 2013). An interesting recent application to social science, particularly to economic geography, can be found in Ikeda and Murota (2014).

As the reader probably remembers (it has already been mentioned), a bifurcation is the structural change in the solution of a differential equation. In this section, only a few local bifurcations are shown. I want to show the explicit expression of this equations since the dynamics of the complete system is reduced to what happens in the neighborhood of the bifurcation point. This fact will be very important for the present work, i.e. to note the conciseness of the model, in terms of the mathematical symbology used. The reduced form of the evolution equation is usually called the "normal form".

This discussion, as well as the ones that follow, do not seek to be a technical story at all, the purpose and the important point is to show the level of information necessary to give a quantitative account of the phenomena studied. This will be of vital importance when connecting the concepts of complexity and emergence.

2.3.2.1 The Saddle-Node Bifurcation

The saddle-node bifurcation appears frequently in dynamical systems of finite dimension (Guckenheimer & Holmes, 2013). In this bifurcation, there are two branches of fixed points on one side of the fork point, and no fixed points on the other side. The normal form that governs this bifurcation can be written as

$$\frac{dx}{dt} = p - x^2.$$

Note the extremely simplicity of the previous expression. Notwithstanding the latter, very complicated patterns can be studied with the use of this model (remember that the compact form written above is valid only in the vicinity of the bifurcation).

For example, solitary waves, called solitons, in highly non-linear physical systems often exhibit this type of bifurcation. The examples also include systems modeled via Boussinesq equations, water wave equation (Buffoni et al., 1996; Yang & Akylas, 1997; Chen, 2000), the Swift-Hohenberg equation in pattern formation (Burke & Knobloch, 2007), the non-linear Schrodinger equation (NLS), equations with localized or periodic potentials in non-linear optics and Bose-condensed systems. Einstein (Herring et al., 2005; Kapitula et al., 2006; Sacchetti, 2009; Akylas et al., 2012), and many others.

2.3.2.2 Transcritical Bifurcation

It is a typical bifurcation that occurs when two stationary solutions change their stationary properties for a critical value of the system. The normal form of this bifurcation is

$$\frac{dx}{dt} = px - x^2,$$

The latter is used extensively to describe, in some cases extremely successfully, the population growth of a given species (May, 1976), whose number represents the variable x. When the value of the number of individuals is sufficiently small, the population grows or dies exponentially according to whether p (the rate of growth, exponential, of the system) is greater or less than zero. If the population grows, after some time, it will have become so large that the shortage of food or the activity of the predators will take effect and the rate of growth will decrease. The nonlinear term in the equation represents just this saturation of the exponential growth of the population.

The stationary points of this system are then: $x = 0$, $x = p$. It is easy to see that there will be a change in stability behavior at these stationary points depending on the sign of p. If it is less than zero, the stable point is $x = 0$, while $x = p$ is unstable. When p is greater than zero, the stability properties of these points change, that is, $x = 0$ is unstable and $x = p$ is stable. The bifurcation point in this case is also $p = 0$.

2.3.2.3 Pitchfork Bifurcation

The case of the pitchfork bifurcation very interesting. It can be associated with some symmetry properties of the system. It is worth remembering that in physical systems, for each continuous mathematical symmetry, there is a corresponding conserved quantity, which is in fact an extremely important property (Noether, 1918), think for example in systems with conservation of energy.

The evolution equation for this bifurcation was written by Landau in 1944, to describe the effect of nonlinearities on hydrodynamic instability. Representing the amplitude of a perturbation by x

$$\frac{dx}{dt} = px \pm x^3.$$

When the sign in the cubic term is negative, the bifurcation is called supercritical. In this case, for p less than zero there is only one equilibrium at $x = 0$. While for p greater than zero there are three solutions, one unstable, at $x = 0$ and two stables at the positive and negative square roots of p. The subcritical case is when the sign of the cubic term, in the evolution equation, is positive. There is an inversion of the stability properties of the solution. For the subcritical case, p less than zero, there are three stationary solutions: $x = 0$ stable, and both unstable solution from the square root solution. For p greater than zero, the only (unstable) solution is $x = 0$. Clearly, the bifurcation point occurs at $p = 0$.

Again, notice the different possibilities of such a simple equation as the pitchfork bifurcation.

2.3.2.4 Pattern Formation

In line with the previously discussed topics, there is a very important phenomenon, formulated mathematically for the first time in the context of morphogenesis. This also corresponds to a type of bifurcation, which in its simplest form is known as the Turing instability (in recognition of the English mathematician Alan Turing that first described it analytically in 1952).

Patterns, or structures (many of them with well-defined geometrical characteristics), appear everywhere in nature and have been one of the important examples cited in the emergence's literature (Bedau & Humphreys, 2008). Spatio-temporal patterns can be observed in chemical reactions or in living systems such as bacterial cultures (Murray, 2002), phenomena involving planetary scales such as the red spot of Jupiter, etc. (note that the analytical understanding can be using different models, according to the particular system).

It is very important to note that, as in almost all quantitative studies, the usual treatment of pattern formation is done by modeling macroscopic interactions (Hoyle, 2006, Cross & Greenside, 2009), since in general the scale or the length of the observed pattern is much greater in orders of magnitude than the size of the microscopic interaction that generates it (I will return to this point later, explaining the agent-based modeling). After discussing the typical mathematical formulation of the Turing mechanism, we will give an important example of the application of this model in the social sciences.

2.3.2.5 Morphogenesis and the Alan Turing's Synthesis

Some of the early ideas and mathematical descriptions on how physical processes and constraints affect biological growth and, therefore, natural patterns such as the spirals of phyllotaxis, were described by D'Arcy Wentworth Thompson in his beautiful book "On Growth and Form" (Thompson, 1917). There, Thompson explained that from his point of view, the forms (the shape) of an animal's body are created by the different speeds of growth in several directions. For example, when the spiral shell of a snail is generated. These mechanisms were quite complicated to understand since they required a very well stipulated, and complicated, internal mechanics to produce the different designs that appear everywhere in nature.

Much later, in the twentieth century, more specifically in 1952, Alan Turing predicted a mechanism for morphogenesis (Turing, 1952). Using only two different chemical signals and the properties of each diffusive process: one that activates and one that inhibits the growth. This type of mechanisms correctly described developmental patterns, and notoriously was only experimentally observed decades after the theoretical work of Turing (Castets et al., 1990). The most complete understanding of the mechanisms involved in real organisms required the discovery of the structure of DNA and the development of molecular biology and biochemistry. Notice that the later involves a different level of description, a different coarse-graining of the theory. The Turing model is considered a valid metaphor for biological morphogenesis, useful for a conceptual framework and to guide modeling, but not for exact prediction in complex bio-physical-chemical systems. However, in very recent works, it is shown that the Turing's model quantitatively explains the morphogenesis in which the cellular compartments first differentiate chemically and then physically (Sheth et al., 2012; Tompkins et al., 2014).

It is important, for the discussion presented in this work, to realize the simple mechanism proposed by Alan Turing. We will describe it succinctly below.

2.3.2.6 The Turing Mechanism

The word diffusion comes from the Latin: *diffudere*, to spread out. Until the work of Turing in 1952, the diffusion was usually thought of as a mechanism that homogenizes the system where it is acting. One of the important insight of Turing's work was to demonstrate how a diffusion mechanism can concentrate elements from a system in a particular region, creating spatio-temporal patterns.

Let's write a classical (complete) reaction-diffusion equation that describes the Turing's instability

$$\frac{\partial u(x,t)}{\partial t} = f(u,v) + Du\frac{\partial^2 u(x,t)}{\partial x^2}$$

$$\frac{\partial v(x,t)}{\partial t} = g(u,v) + Dv\frac{\partial^2 v(x,t)}{\partial x^2}$$

I keep the discussion simple, and guide the curious readers to a more formal and complete discussion on the subject in the references below (Murray, 2002; Cross & Greenside, 2009).

In the previous equations, we have the reaction parts: the functions f, and g; and the diffusion terms, characterized by two diffusion parameters Du and Dv. A necessary condition for the Turing instability is a difference in these diffusion coefficients, in particular the diffusion of the so called inhibitor, v, must be greater that the activator u. With the right conditions, there is a bifurcation point where the solution of this equation changes from a uniform solution to a patterned one.

The successful use of this type of theory to model emergent phenomena is remarkable. In a recent paper by Lim, Metzler, and Bar-Yam, the authors have studied the formation of global social patterns and ethnic-cultural violence (Lim et al., 2007). Using a somewhat more sophisticated model than the one previously discussed, they predict the emergence of conflict zones, i.e. the patterns, in Eastern Europe with amazing precision. This type of study shows the power and universality of these concepts and the variety of applications of this methodology to different fields of study. In terms of information, very few bits are needed in the model to account for the phenomenon studied.

2.3.3 *Network Theory*

Until now I have dealt with continuous models. The case of a network is different. A network is a set of nodes (also called vertices) that are connected by edges (Newman et al., 2011). Network theory can be traced back to the celebrated Konigsberg Bridge problem and its solution by Euler in 1736, which has been treated as the formal beginning of graph theory, a mathematical theory that preceded network theory. It is very important to realize that there are many instances where approaches of the continuous type, like the Turing mechanism (see equations above), are not a good approximation for the problem in hands and only a network view will accurately describe the phenomena.

Among this type of systems (which are usually cited in the emergence's literature) are social networks, with connections between individuals; transport networks in cities or between cities, as airplane networks; the World Wide Web; food webs; neuronal networks; collaboration networks: scientific, organizations or business communities; etc.

Some important concepts must be understood when studying networks since many of them not only characterize topologically, but they shed light on the mechanisms that make global emergence attributes. Some are (Costa et al., 2007; Newman, 2010; Gross et al., 2006)

- Degree: Is the number of vertices connected to a node. One can easily see the importance of this, but after some recent works, we now know that in the case of many systems this is not the most important concept since, for example, nodes that are poorly connected act only as of the intermediaries between parts of a big network.
- Directed or undirected: If an edge runs only in one direction, it is called direct, think for example in a one-way road. In contrast, if the edge goes in both directions, it is called undirected.
- Geodesic Path: This is the shortest path, following the connecting edges, between two nodes. This quantity is important when addressing questions such as propagation of information, the spread of infectious diseases, or alike.

2.3.4 Game Theory

The emergence of systems with the ability to generate strategies during their evolution is one of the fascinating topics in this area. This is what Bedau and Humphreys (2008) suggest: "In addition, the evolutionary processes that shape biological lineages imply emergence. A complex and highly differentiated biosphere has emerged over billions of years from what was originally a much simpler and much more uniform variety of early life forms."

Game theory, or the theory of social dilemmas, focuses on how a group of elements interacts using strategic decision making. Although the history of game theory can be traced back to the early 1700s, the modern version of it appears after the work of John von Neumann in 1928 (von Neumann & Morgenstern, 2007).

Several works follow von Newmann's efforts, I can mention the important work made by Nash in 1950a, 1950b, which introduced the idea of Nash Equilibrium, a mutual consistency of strategies. Nowadays this theory is also applied in different fields such as political science, biology, economics, computer science, etc. An interesting example of a branch of this discipline is evolutionary theory, which focuses on the dynamics of the strategy change. In this context, games are called evolutionary games.

In the classical set up of game theory, the players have movement choices (decisions they can make with different payoffs), and the game can be in a single round or repetitive ones. The rules or choices the players can make are usually arranged in decision trees or matrixes.

Let's show a very simple example to see how the theory works, I will discuss a well-known example in game theory: the prisoner dilemma. Two players are partners in a crime and after been captured, on suspicion of its commission, they are confined in different cells. The police offer them the opportunity to confess the crime. We can then represent the players in a two by two matrix with the different pay-offs of the four possible choices depending on the criminal confessions:

- Prisoner A stays silent, Prisoner B stays silent: Each serves 1 year;
- Prisoner A stays silent, Prisoner B betrays: Prisoner A: 3 years, Prisoner B: is released;
- Prisoner A betrays, Prisoner B stays silent: Prisoner A: is released, Prisoner B: 3 years;
- Prisoner A betrays, Prisoner B betrays: Each serves 2 years.

The best possible outcome for the two prisoners is to not confess. If only one confesses, she/he gains a lot of utility while the other loses. The alternative is the confession of the two prisoners. What would then be the most probable outcome in this scenario?

As the reader can guess, game theory can be applied also in complex networks, taking into account the topology where the individual interacts: the social network. There are many types of games, we can mention: cooperative, non-cooperative; discrete and continuous games; simultaneous, sequential; evolutionary games; perfect or imperfect information; many players, population games, etc.

It is important to realize that in all the situations, for the particular case of rational behavior, the players (which can be a person, a biological agent, etc.) must anticipate what to do, taking into account what the other agent will infer from the other's actions (Camerer, 2003).

The results at the macroscopic level of different strategies, for many agents systems, are considered an emerging phenomenon in the literature (Risjord, 2014).

2.3.5 *Information Theory*

Claude Shannon developed a theory to find the limits of signal processing; his work "A Mathematical Theory of Communication" was published in the Bell System Technical Journal in July and October 1948. This is the landmark for what now is called Information Theory (Shannon, 1948). Since Shannon's work, Information Theory has been successfully applied to different fields (J. R. Pierce, 1980; Cover & Thomas, 2012): molecular genetics, cryptography, statistical inference, physics, biology, and in general to data analysis. In complex systems, information theory has been used in connection with a theory that was developed by E.T. Jaynes. In a series of papers circa 1952, he discussed the correspondence between statistical mechanics and information theory (Rosenkrantz, 1983). This giant step tells us that statistical mechanics (and all the applications/predictions of this, very successful, body of knowledge) must be seen as a particular case of a more general theory: Information Theory. Jaynes's work paid attention to a general principle: the Maximum Entropy Principle (or MaxEnt). Nowadays MaxEnt is used to understand several distributions appearing in biology and ecology (from a complex system approach) as size distribution, range distribution, energy distribution, etc. In a very recent effort, I applied the MaxEnt theory to understand social and urban systems, but this work is still unpublished and in progress.

2.3.6 Super-Statistic

I have emphasized the importance of the contribution of statistical mechanics and MaxEnt theory to the study of complex systems and natural phenomena in general. In recent work, Cohen et al. introduced a natural generalization of statistical mechanics (Beck & Cohen, 2003; Cohen, 2004). The idea is very simple, but also very powerful. When dealing with complex non-equilibrium systems with long term stationary states subject to spatiotemporal fluctuations of an intensive quantity, the probability distribution, which has very peculiar characteristics, can be obtained by calculating the average over these fluctuations. A typical example of these characteristics may be the so-called behavior of distributions of extreme events (such as earthquakes, fluctuations, and falls of financial exchanges, etc.), usually characterized by scalable functions such as those mentioned at the beginning of this chapter. This characteristic is present in a variety of emerging phenomena, for example, those associated with critical self-organization (Zelinka et al., 2013).

To be more explicit, let's look at an example to see how the theory works. Suppose we have a system composed of many subsystems. Each subsystem has particles that diffuse at an average speed well characterized by a diffusion parameter. Consequently, each subsystem will have a well-defined distribution characterized, as mentioned, by its particular diffusion parameter. But, if we consider the complete system (the aggregation of all subsystems) we must average using all these particular distributions. When doing so, new emergence characteristics appear.

It is easy to realize the power of this concept and the generalization that can be made following these methods (Hanel et al., 2011).

When considering a compound of elements, it is necessary to first check if their interaction can be decomposed into subsystems. There is no formal method to do this, the only tool to know what is the proper subdivision of the system is the information in hand of the problem. For example, in a recent paper, financial time series data was divided in days, i.e. a trading-day was the temporal division for the complete system, which was many years of trading-days (Gerig et al., 2009). If this sort of subdivision can be made, these subsystems will follow the same underlying dynamics (as the Brownian particle in the example above), and the only difference will be some fluctuation in an intensive variable. Then, the behavior of the aggregated system will simply be an average of the subsystems.

2.3.7 Cellular Automata

Circa 1950 Stanislav Ulam and John von Neumann created a model to understand the behavior of discrete units as a function of the behavior of its neighboring ones. It was the beginning of Cellular Automaton models. Cellular Automata is a discrete model based on cells, each one having a set of states: on–off or alike. The cell positions are usually on a regular grid (but again they can be arranged in complex networks like the one mentioned before). Then, given an initial condition for the cellular automata, the next state will be an update of each grid according to local rules (Toffoli & Margolus, 1987; Schiff, 2011).

To explain the essential ideas, let's take a look at a simple example: the one-dimensional cellular automata. In a one dimensional cellular automata, each cell can be in two states: zero and one (or on and off, etc.). Given the state of a cell at time t, its configuration at time t + 1 will depend on its state at time t and the state of the two neighbors also at time t. It is clear then that the possible values for a neighborhood are two to the power of three, i.e. $2^3 = 8$, and then given the on or off options, there will be a total of $2^8 = 256$ rules for a one-dimensional cellular automata.

Depending on their behavior S. Wolfram, in his book "A New Kind of Science", defined four categories into which cellular automata can be classified. In class one nearly all initial patterns evolve quickly into a stable, homogeneous state, any randomness in the initial pattern disappears. In class two nearly all initial patterns evolve quickly into stable or oscillating structures, some of the randomnesses in the initial pattern may filter out, but some remain. Local changes to the initial pattern tend to remain local. Class three has nearly all initial patterns evolving in a pseudo-random or chaotic manner. Any stable structure that appears is quickly destroyed by the surrounding noise. Local changes to the initial pattern tend to spread indefinitely. Finally, in class four nearly all initial patterns evolve into structures that interact in complex and interesting ways, with the formation of local structures that can survive for long periods.

Applications of Cellular Automata range can be used in different fields such as computer processors used to understand pattern formation in biology, epidemiology, models to simulate urban dynamics through the local actions of cellular automata (Batty, 2007), etc. They have also been the paradigmatic example and widely discussed in emergence (Gardner, 1970; Goldstein, 1999; Chalmers, 2006a, 2006b; Bedau, 2008; Beisbart, 2012, Frigg & Reiss, 2009) using some rules in particular, such as the so-called "Game of life" by the English mathematician J. H. Conway.

2.3.8 *Agent Base Modelling*

With the arrival of new technologies and the increasing of computational power, it is straightforward to consider computational models to study the evolution of many agents, at different scales and scenarios. Agent-based modeling can be thought of as the evolution of cellular automata models. This can be considered as a bottom-up approach since the properties observed in the system as a whole (i.e. emergent properties) are the result of the interactions of the microscopic components of the system. Such view differs from the others discussed in this chapter, for example, the Turing mechanism, where the diffusion of particles is modeled through a spatial operator that acts on a macroscopic scale.

There is no specific recipe to apply agent-based modeling since it can be used in many different frameworks and systems in general. Usually, they can be studied at various levels, such as individuals (Axelrod, 1997), population (Gustafsson & Sternad, 2010), organizations, etc.; models for decision making (notice that game theory can be applied as well); the topology of interactions, regular or irregular lattices, complex networks; the environment where the interaction happens and learning rules (or adaptive processes).

It has been argued that the main benefits of agent-based modeling are the following (Bonabeau, 2002):

- It captures emergent phenomena, this is because in principle emergent phenomena come from microscopic interactions (or individual entities), and when using agent-based models any macroscopic characteristic will be by definition a result of microscopic rules acting on a great number of agents.
- This emergent phenomenon can appear when: individual behavior is nonlinear, agent interactions are heterogeneous and can generate network effects, agents-based models can amplify fluctuations (something that is difficult to achieve aggregating differential equations) and when individual behavior exhibits path dependence and/or memory (like in the case of a traffic jam).
- In many cases agent-based modeling provides the most natural way to describe dynamics and rules of the system, focusing on the individual rules of the agents.

Some of the areas where agent-based modeling is being applied are: diffusion of innovation and adoption, operational risk, organization design, stock market, flows (as for example traffic or evacuation), etc.

2.3.9 *Data Mining*

There are many systems where the underlying dynamics are unknown. These are in some way very different forms of the classical type of dynamical systems, where first-principle models can be proposed to describe them. In these systems, Newton's equation of motion, for classical systems, Maxwell equations, for electromagnetic systems, quantum or relativistic equations, can be the starting point to build models that will describe the phenomenon using a bottom-up approach. For other systems this type of reduction is impossible due to: the complexity of the problem, which sometimes makes it almost impossible to create a mathematical model; the lack of information since sometimes the data, or output, of the system, is available only at a very high and macroscopical level; etc.

Nowadays there is much data available: from scientific institutions, governments, different types of businesses, the world wide web, etc. All this available data can be stored and studied, but as the reader can guess for some of them the first-principle approach is far from being achieved.

The term data mining does not refer to the extraction and collection of a huge amount of data, as usually is believed. Data mining refers to the extraction and recognition of patterns in large data sets. In this sense, data mining has two primary goals (Kantardzic, 2011): prediction and description. To achieve them, data mining uses the following task: classification, regression, clustering, summarization, modeling, and deviation detection.

Even though the recognition of patterns and sometimes the detection of signs of causality in some interactions has been a research topic using old methods (e.g. correlation and regression analysis); data mining explores a huge amount of data available thanks to the increment of computer power and storage. Data mining identifies unknown (patterns) properties present in the data, by using (among others) artificial intelligence and machine learning techniques (Hastie et al., 2009). Due to the variety of data available nowadays, thanks to the internet, cellphone communications, etc., data mining is becoming a powerful tool to study social patterns in urban systems. Common applications of this branch of computer science can be seen in astronomy, genetics, social behavior, transport, financial systems, telecommunication, etc.

2.4 Complexity Measures

We have just summarized the variety of methodologies with which a complexity science's scientist works when studying a system that (surely according to his perception) can be classified as complex. At the same time, much has been written trying to define (or measure) the complexity of a system. The importance of this can be understood when thinking that systems with the same level of complexity (defined in some way) may share universal properties, i.e. systems with the same level of complexity (having chosen the same definition of the measure of complexity) could share characteristics. Which undoubtedly is of epistemological importance.

The history of the study of probabilistic regularities in physical systems can be traced back to 1857 with the very idea of entropy proposed by Rudolf Clausius. Claude Shannon derived the same functional form used almost one century before introducing the concept of information entropy. Even though the intuitive idea of complexity and information in a physical system share some similarities, it was necessary to introduce several measures to understand various types of complexities, and to quantify properties of the system closely related with both of them (Lloyd, 2001): Kolmogorov complexity, logical depth, effective complexity, etc.

Some of the different proposed measures of complexity are associated with the study of particular systems, for example, the study of strings of symbols, or the data that the system produces in a spatio-temporal manner, etc. Following then Lloyd (2001) we can group certain types of questions related to the concept of complexity in a given system, these can be:

- How difficult is it to describe?
- How hard is it to create?
- What is the degree of organization?

We will see then that each of these questions leads us to deep philosophical questions that involve at least epistemological aspects (ability to reduce the observer, theories that describe the phenomenon, levels of organization, etc.). Next, we will discuss some of these measures of complexity, noting that the technically detailed study can be found in the mentioned references, we will analyze here what is relevant for this discussion.

2.4.1 Computational Complexity

The theory of computational complexity basically focuses on classifying computational problems according to their inherent difficulty, which can be: time, that is, the complexity that exists in terms of the execution time that resolves it; spatial complexity, in terms of the memory that is needed for the calculation; or in terms of quantity and type of mathematical operations (Arora & Barak, 2009).

In this theory it is understood that a computational problem is a task that in principle is capable of being solved by a computer, which is to say that the problem can be solved by a mechanical application, usually linear, of mathematical steps, such as an algorithm.

To graph these points we can mention a classic problem in this area: the travelling salesman problem (TSP), which raises the following question: given a list of cities and their geographical points, which is the shortest route possible to visit, taking note that each city is visited exactly once, and that the seller returns to the origin city? This is a very difficult problem in terms of combinatorial optimization, and very important in operational research and theoretical informatics.

Despite its long history (Laporte, 2006), which dates back to the works of Euler around 1736, the problem was rigorously formulated for the first time in 1930 and is one of the most intensively studied problems in optimization theory. It is used as a reference point for many methods. Although the problem is computationally difficult, a large number of heuristics and exact algorithms are known, so that some cases with tens of thousands of cities can be solved.

The TSP has several applications, even in its purest form, such as planning, logistics and mobility in social systems, etc. Slightly modified, it appears as a subproblem in many areas, such as DNA sequencing. In many applications, additional limitations may be imposed on the classical problem, such as limited resources or maximum city-to-city times.

So, in this case, a problem is considered inherently complex if its solution requires significant resources (in at least the three types of difficulties mentioned above), whatever the algorithm used. The theory formalizes this intuition, introducing mathematical models to study these problems and quantify the amount of resources necessary to solve them. Also, more specific complexity measures are also used, such as the amount of communication (used in complexity of communication), the number of gates in a circuit or the number of processors used.

One of the roles of computational complexity theory is to determine the practical limits of what computers can and cannot do.

Closely related fields in theoretical informatics are the analysis of algorithms and the theory of computability. A key distinction between the analysis of algorithms and the theory of computational complexity is that the first is devoted to analyzing the amount of resources needed to solve a problem, while the last one asks a more general question about all the possible algorithms that could be used to solve the same problem. More precisely, the theory of computational complexity attempts to classify problems that can or cannot be solved with adequately restricted resources. In turn, imposing restrictions on available resources is what distinguishes the computational complexity of computability theory: this last theory asks what kind of problems can, in principle, be solved algorithmically.

2.4.2 Algorithmic Complexity

The Kolmogorov complexity, *K*, was introduced independently, in the 1960s by Solomonoff, Kolmogorov and Chaitin (Solomonoff, 1964; Kolmogorov, 1963, 1965; Chaitin, 1966).

Formally, *K*, laid within the algorithmic theory of information, in the area of information technology, which studies not only the complexity of Kolmogorov but also other measures of complexity applied to vectors or data chains (in general they can be other data structure).

Simply speaking, the Kolmogorov complexity of an object, reduced to a data vector, is the length of the shortest computer program (in a given programming language) that produces the object. It is a measure of the computational resources required to specify the object and is also known as algorithmic, descriptive complexity, Kolmogorov-Chaitin complexity, algorithmic entropy or program size complexity. There are several variants of the Kolmogorov complexity or algorithmic information. The most commonly used type of complexity K is based on self-delimitation programs and is mainly due to Leonid Levin (1974).

Despite its usefulness, Kolmogorov's complexity does not capture very well the intuitive notion of "how complex and object is". For example, as it may be obvious to the reader, a random data vector without any regularity, have a very large Kolmogorov Complexity. But those string of data are not "complex" from an intuitive point of view, they are completely random and have no interesting structure (or regularities) at all.

2.4.3 *Effective Complexity*

I think the one that captures in a more accurate way the notion of the complexity of a given system is Effective Complexity, introduced by Murray Gell–Mann and Seth Lloyd (Gell-Mann, 1995; Gell-Mann & Lloyd, 1996). In short, the effective complexity of an entity corresponds to "the length of a highly compressed description of its regularities".

The idea is simple, elegant, and profound: if we split the algorithmic information content of some data string into two components: one with its regularities (related to the Kolmogorov Complexity) and the other with its random features (related to its entropy), the Effective Complexity of the data will only be the algorithmic information content of its regularities.

This definition has clear epistemological consequences. A perceptive reader will notice a very important aspect of the theory developed by Gell–Mann, and Lloyd, that the effective complexity of an entity is context-dependent (Gell-Mann & Mermin, 1994). We will give a naive example to motivate the analysis of this aspect of the theory. Imagine we are studying a particular system: a living organism, then, what is its complexity?

There is no doubt that we must be more specific mentioning exactly which characteristic or feature we want to study using this concept, and what set of data we have to do so. Not only that but to be more precise, we must have a theory that explains the data.

2.5 Theories and Models

In the field of Philosophy of Science, the distinction on what can be considered theory has been a topic of great discussion from the beginning of the last century and is still a subject of analysis today. Similarly, in the second half of the same century, the concept of "model" entered into the focus of attention of the philosophical community, in a certain sense in competition with the theory concept, by fulfilling some of its functions and distinguishing itself by avoiding some of the discussions about the commitments that the use of terms could imply in a theory, for which its reference is doubtful having been introduced into the conjectural process.

Thus it would be impossible to review here without committing serious omissions, the different positions that attempt to account for theories and models. It must, however, be borne in mind that for the predictions that can be taken as an index of the success of a conceptual scheme that pretends to account for a collection of data, both types of entities, theories and models face the same challenges.

The degree of success in predicting experimental results is an indicator that the theory has not been refuted, has been corroborated, is a plausible conjecture, seems appropriate for that range of phenomena for which it has been proposed, and a series of other impressions that reinforce confidence in the use of that theory.

Similarly, the same success in predicting the results is taken as an indication that the model used is an appropriate one, it has been constructed in such a way that its use can be guaranteed to justify intervention actions in the environment, it seems even possible that it captures part of the underlying processes in the relation between inputs and outputs, and all other series of affirmations that also reinforce the persistence in the use of that model for that range of phenomena.

Thus, models and theories seem to be structures that obtain feedback from the data that stimulate or discourage their choice to account for these or those phenomena. It is in this sense that both notions can be used in a similar way even when all other extensions of the similarity can run serious risks. As a sample of this, it is worth noting that while some models that allow the calculation of results for certain experimental configurations can be constructed knowing their lack of correspondence with the real processes that take place in the nature to which they are trying to apply, this would not be a typical construction for a (physical) theory: because in it is always present the expectation of description of the entities and processes that take place in the phenomenological world.

Regarding our objectives, it will not be necessary to distinguish between theories and models when discussing the notions of complexity and their use when accounting for a given phenomenon. For this reason, both models and theories are covered in the conclusions that are obtained in this investigation, and, in each case that the term "theory" has been written could have also appeared "model", and vice versa.

References

Akylas, T. R., Hwang, G., & Yang, J. (2012). From non-local gap solitary waves to bound states in periodic media. In *Proceedings of the Royal Society A* (Vol. 468, pp. 116–135). The Royal Society.

Arora, S., & Barak, B. (2009). *Computational complexity: A modern approach*. Cambridge University Press.

Axelrod, R. M. (1997). *The complexity of cooperation: Agent-based models of competition and collaboration*. Princeton University Press.

Batty, M. (2007). *Cities and complexity: Understanding cities with cellular automata, agent-based models, and fractals*. The MIT Press.

Beck, C., & Cohen, E. G. D. (2003). Superstatistics. *Physica A: Statistical mechanics and its applications, 322*, 267–275.

Bedau, M. A. (2008). Is weak emergence just in the mind? *Minds and Machines, 18*(4), 443–459.

Bedau, M. A., & Humphreys, P. E. (2008). *Emergence: Contemporary readings in philosophy and science*. MIT Press.

Beisbart, C. (2012). How can computer simulations produce new knowledge? *European Journal for Philosophy of Science, 2*(3), 395–434.

Boccara, N. (2010). *Modeling complex systems*. Springer Science & Business Media.

Bonabeau, E. (2002). Agent-based modeling: Methods and techniques for simulating human systems. *Proceedings of the National Academy of Sciences, 99*(suppl 3), 7280–7287.

Buffoni, B., Champneys, A. R., & Toland, J. F. (1996). Bifurcation and coalescence of a plethora of homoclinic orbits for a Hamiltonian system. *Journal of Dynamics and Differential Equations, 8*(2), 221–279.

Burke, J., & Knobloch, E. (2007). Homoclinic snaking: Structure and stability. Chaos: An interdisciplinary. *Journal of Nonlinear Science, 17*(3), 037102.

Camerer, C. (2003). *Behavioral game theory: Experiments in strategic interaction*. Princeton University Press.

Castets, V., Dulos, E., Boissonade, J., & De Kepper, P. (1990). Experimental evidence of a sustained standing Turing-type nonequilibrium chemical pattern. *Physical Review Letters, 64*(24), 2953–2956.

Chaitin, G. J. (1966). On the length of programs for computing finite binary sequences. *Journal of the ACM (JACM), 13*(4), 547–569.

Chalmers, D. J. (2006a). *The re-emergence of emergence. Chapter strong and weak emergence*. Oxford University Press.

Chalmers, D. J. (2006b). *Strong and weak emergence. The reemergence of emergence* (pp. 244–256). Oxford University Press.

Chen, M. (2000). Solitary-wave and multi-pulsed traveling-wave solutions of Boussinesq systems. *Applicable Analysis, 75*(1–2), 213–240.

Cohen, E. G. D. (2004). Superstatistics. *Physica D: Nonlinear Phenomena, 193*(1), 35–52.

Costa, L. D. F., Rodrigues, F. A., Travieso, G., & Villas Boas, P. R. (2007). Characterization of complex networks: A survey of measurements. *Advances in Physics, 56*(1), 167–242.

Cover, T. M., & Thomas, J. A. (2012). *Elements of information theory*. Wiley.

Cross, M., & Greenside, H. (2009). *Pattern formation and dynamics in nonequilibrium systems*. Cambridge University Press.

Euler, L. (1736). Solutio problematis ad geometriam situs pertinentis. *Commentarii Academiae Scientiarum Imperialis Petropolitanae, 8*, 128–140.

Feynman, R. P., Leighton, R. B., & Sands, M. L. (1964). *The Feynman lectures on physics: Electromagnetism and matter* (Vol. 2). Addison Wesley Publishing Company.

Frigg, R., & Reiss, J. (2009). The philosophy of simulation: Hot new issues or same old stew? *Synthese, 169*(3), 593–613.

Gardner, M. (1970). Mathematical games: The fantastic combinations of John Conway's new solitaire game life. *Scientific American, 223*(4), 120–123.

Gell-Mann, M. (1995). What is complexity? *Complexity, 1*(1), 16–19.

Gell-Mann, M., & Lloyd, S. (1996). Information measures, effective complexity, and total information. *Complexity, 2*(1), 44–52.

Gell-Mann, M., & Mermin, D. (1994). The quark and the jaguar: Adventures in the simple and the complex. *Physics Today, 47*, 89.

Gerig, A., Vicente, J., & Fuentes, M. A. (2009). Model for non-Gaussian intraday stock returns. *Physical Review E, 80*(6), 065102.

Goldstein, J. (1999). Emergence as a construct: History and issues. *Emergence, 1*(1), 49–72.

Gross, T., Lima, C. J. D., & Blasius, B. (2006). Epidemic dynamics on an adaptive network. *Physical Review Letters, 96*(20), 208701.

Guckenheimer, J., & Holmes, P. J. (2013). *Nonlinear oscillations, dynamical systems, and bifurcations of vector fields* (Vol. 42). Springer Science & Business Media.

Gustafsson, L., & Sternad, M. (2010). Consistent micro, macro and state-based population modelling. *Mathematical Biosciences, 225*(2), 94–107.

Hale, J. K., & Koçak, H. (2012). *Dynamics and bifurcations* (Vol. 3). Springer Science & Business Media.

Hanel, R., Thurner, S., & Gell-Mann, M. (2011). Generalized entropies and the transformation group of superstatistics. *Proceedings of the National Academy of Sciences, 108*(16), 6390–6394.

Hastie, T., Tibshirani, R., & Friedman, J. (2009). The elements of statistical learning: Data mining, inference, and prediction. *Biometrics*.

Herring, G., Kevrekidis, P. G., Carretero-González, R., Malomed, B. A., Frantzeskakis, D. J., & Bishop, A. R. (2005). Trapped bright matter-wave solitons in the presence of localized inhomogeneities. *Physics Letters A, 345*(1), 144–153.

Holland, J. H. (1998). *Emergence: From chaos to order*. Oxford University Press.

Hoyle, R. B. (2006). *Pattern formation: An introduction to methods*. Cambridge University Press.

Ikeda, K., & Murota, K. (2014). *Bifurcation theory for hexagonal agglomeration in economic geography*. Springer.

Kantardzic, M. (2011). *Data mining: Concepts, models, methods, and algorithms*. Wiley.

Kapitula, T., Kevrekidis, P. G., & Chen, Z. (2006). Three is a crowd: Solitary waves in photorefractive media with three potential wells. *SIAM Journal on Applied Dynamical Systems, 5*(4), 598–633.

Kolmogorov, A. N. (1963). On tables of random numbers. Sankhy? *The Indian Journal of Statistics, Series A*, 369–376.

Kolmogorov, A. N. (1965). Three approaches to the quantitative definition of information. *Problems of Information Transmission, 1*(1), 1–7.

Laplace, P. S. (1825). Pierre-Simon Laplace philosophical essay on probabilities: Translated Leibniz from the fifth French edition of 1825 with notes by the translator (Vol. 13). Springer Science & Business Media, 2012.

Laporte, G. (2006). A short history of the traveling salesman problem. Canada Research Chair in Distribution Management, Centre for Research on Transportation (CRT) and GERAD HEC Montréal, Canada.

Laudan, L. (1986). *El progreso y sus problemas. Hacia una teoría del crecimiento científico*. Encuentro Ediciones.

Levin, L. A. (1974). Laws of information conservation (nongrowth) and aspects of the foundation of probability theory. *Problemy Peredachi Informatsii, 10*(3), 30–35.

Lim, M., Metzler, R., & Bar-Yam, Y. (2007). Global pattern formation and ethnic/cultural violence. *Science, 317*(5844), 1540–1544.

Lloyd, S. (2001). Measures of complexity: A nonexhaustive list. *IEEE Control Systems Magazine, 21*(4), 7–8.

May, R. M. (1976). Simple mathematical models with very complicated dynamics. *Nature, 261*(5560), 459–467.

Miller, J. H., & Page, S. E. (2007). Complexity in social worlds, from complex adaptive systems: An introduction to computational models of social life. Introductory Chapters.

Murray, J. D. (2002). *Mathematical biology. I & II, volume 17 of interdisciplinary applied mathematics*. Springer.

Nash, J. F. (1950a). The bargaining problem. *Econometrica: Journal of the Econometric Society, 18*, 155–162.

Nash, J. F. (1950b). Equilibrium points in n-person games. *Proceedings of the National Academy of Sciences, 36*(1), 48–49.

Newman, M. (2010). *Networks: An introduction*. Oxford University Press, Oxford.

Newman, M., Barabasi, A. L., & Watts, D. J. (2011). *The structure and dynamics of networks*. Princeton University Press.

Nicolis, G., & Nicolis, C. (2012). *Foundations of complex systems: Emergence, information and prediction*. World Scientific.

Noether, E. (1918). Invariante Variationsprobleme Nachr. D. König. Gesellsch. D. Wiss. Zu Göttingen, Math-phys. Klasse 1918: (pp. 235–257).

Pierce, J. R. (1980). An introduction to information theory: Symbols. *Signals and Noise E, 2*.

Poincaré, H. (1885). Sur l'équilibre d'une masse fluide animée d'un mouvement de rotation. *Acta Mathematica, 7*(1), 259–380.
Risjord, M. (2014). *Philosophy of social science: A contemporary introduction*. Routledge.
Rosenkrantz, R. D. (Ed.). (1983). *ET Jaynes: Papers on probability, statistics and statistical physics* (Vol. 158). Springer Science & Business Media.
Sacchetti, A. (2009). Universal critical power for nonlinear Schrödinger equations with a symmetric double well potential. *Physical Review Letters, 103*(19), 194101.
Scheffer, M., Bascompte, J., Brock, W. A., Brovkin, V., Carpenter, S. R., Dakos, V., et al. (2009). Early-warning signals for critical transitions. *Nature, 461*(7260), 53–59.
Schiff, J. L. (2011). *Cellular automata: A discrete view of the world* (Vol. 45). Wiley.
Shannon, C. E. (1948). A mathematical theory of communication, part I, part II. *Bell System Technical Journal, 27*, 623–656.
Sheth, R., Marcon, L., Bastida, M. F., Junco, M., Quintana, L., Dahn, R., et al. (2012). Hox genes regulate digit patterning by controlling the wavelength of a Turing-type mechanism. *Science, 338*(6113), 1476–1480.
Solomonoff, R. J. (1964). A formal theory of inductive inference. Part I. *Information and Control, 7*(1), 1–22.
Strogatz, S. H. (2014). *Nonlinear dynamics and chaos: With applications to physics, biology, chemistry, and engineering*. Westview Press.
Thompson, D. (1917). On growth and form. New York: Mac Millan (published originally by Cambridge University Press, 1917).
Toffoli, T., & Margolus, N. (1987). *Cellular automata machines: A new environment for modeling*. MIT Press.
Tompkins, N., Li, N., Girabawe, C., Heymann, M., Ermentrout, G. B., Epstein, I. R., & Fraden, S. (2014). Testing Turing's theory of morphogenesis in chemical cells. *Proceedings of the National Academy of Sciences, 111*(12), 4397–4402.
Turing, A. M. (1952). The chemical basis of morphogenesis. *Philosophical Transactions of the Royal Society of London B: Biological Sciences, 237*(641), 37–72.
von Neumann, J., & Morgenstern, O. (2007). *Theory of games and economic behavior*. Princeton University Press.
West, G. B., Brown, J. H., & Enquist, B. J. (1997). A general model for the origin of allometric scaling laws in biology. *Science, 276*(5309), 122–126.
Yang, T. S., & Akylas, T. R. (1997). On asymmetric gravity capillary solitary waves. *Journal of Fluid Mechanics, 330*, 215–232.
Zelinka, I., Sanayei, A., Zenil, H., & Rössler, O. (Eds.). (2013). *How nature works: Complexity in interdisciplinary research and applications* (Vol. 5). Springer Science & Business Media.

Chapter 3
Complexity and Emergence

3.1 States of a System

As discussed in previous chapters, the notion of complexity is deeply tied to the interactions among basal elements, which give rise to a diverse range of intriguing and often unexpected emergent phenomena. A defining feature of these phenomena is the multiplicity of possible final states that a system can exhibit. This diversity of outcomes stems from the presence of different levels of interaction and modeling, frequently referred to as "coarse-grained modeling" (Levitt & Warshel, 1975; Ingólfsson et al., 2014).

In contrast to elementary physical phenomena—such as the Newtonian free fall of an object under gravity in a two-body system—complex systems exhibit a variety of potential solutions. These are the product of inherent non-linearities emerging from interactions among entities at a certain (typically lower) level of organization, from which emergent properties arise. As a consequence, such systems possess the capacity to switch between multiple possible states—not only due to thermal fluctuations, but also as a response to external conditions. This capacity enables the system to explore, adapt, and, more generally, evolve.

From a philosophical standpoint, this opens up compelling discussions about causation, co-evolution, and what Peter Godfrey-Smith and others have called "niche construction" (Godfrey-Smith, 2000; Laland et al., 2014), although such topics lie beyond the scope of this work.

Emergence in complex systems can manifest in at least two distinct ways. One concerns the appearance of global properties within a system composed of many agents or components (Kilicay & Dagli, 2003). These global properties emerge from internal dynamics (including fluctuations) and external influences (e.g., boundary conditions, energy or matter exchange), and are irreducible to the properties of individual components. They are often described as "unexpected" or emergent phenomena. Because of this irreducibility, the emergence of new features is typically

M. Fuentes, *Complexity, Emergence and the Evolution of Scientific Theories: Towards a Predictive Epistemology*, SpringerBriefs in Philosophy,
https://doi.org/10.1007/978-3-031-99994-9_3

associated with the formation and maintenance of hierarchical structures. In these structures, disorder and randomness at local levels are, in some way, regulated by higher levels of organization. Macroscopic coherence may also arise from long-range, nonlinear interactions—both temporal and spatial—leading to system-wide states of order and structure.

In contemporary literature, this kind of behavior is frequently discussed under the concept of self-organization. Accordingly, the study of dynamical systems composed of many interacting parts reveals a gradual departure from reductionist approaches, which aim to infer the whole from its parts, toward emergentist perspectives. These perspectives focus on self-organizing phenomena that result from the system's potential to occupy multiple macro-states, depending on its boundary and initial conditions. As Gallagher (2012) notes: "... a new emphasis on neuroscience, and connectionism, which challenged the prevailing computational orthodoxy by introducing an approach based on nonlinear dynamic systems. With this formulation, there was a shift away from an emphasis on reductionism, at least in some circles, to an emphasis on the notion of emergence and self-organization." Classical examples of such behavior can be found in biological systems.

From a historical and philosophical point of view, an extensive body of literature further enriches the discussion and understanding of complex systems. A particularly influential reference is Campbell's seminal work (1974), which provides foundational insights into hierarchical complexity. It is also worth emphasizing that non-linear effects are ubiquitous across observational scales. Many macroscopic physico-chemical phenomena are driven by feedback loops, in which the outcome of a process modifies the conditions under which the process unfolds. These loops—whether positive or negative—play a crucial role in shaping the system's dynamics.

Another mode in which complexity and emergent behavior manifest is through the entanglement of large-scale regularities with dynamic elements that evolve stochastically. These seemingly erratic evolutionary trajectories, such as those discussed by Ohno (1973), are also a hallmark of complex systems.

The coexistence of order and disorder in such systems often gives the impression that the process is out of control—not just temporally, but spatially as well. This raises fundamental epistemological questions about the predictability of long-term behavior. Classical (non-quantum) examples of this unpredictability are found in systems exhibiting spatiotemporal chaos (Agazzi & Montecucco, 2002).

An attentive reader will recognize that if the low-level interactions among the system's components were to produce simple proportional (i.e., linear) relationships, then the system would lack the multiplicity of possible states and behaviors we associate with complexity. Non-linearity is, therefore, a necessary condition for complexity. In this sense, nonlinear science offers a natural framework to describe, analyze, and classify the properties and (auto)evolutionary scenarios of complex systems.

3.2 Transitions Between States

As we have observed throughout this study, a key aspect in the analysis of complex systems lies in understanding transitions between states. Complexity in an evolving system cannot increase if the system remains confined to a single, static state. To explore how such transitions occur, we introduce the notion of a control parameter (Schöll & Schuster, 2008), which describes the various ways in which a system interacts with and is influenced by its surrounding environment.

A straightforward example is a chemical system within a container, where temperature acts as the control parameter. Depending on its value, chemical reactions will proceed at different rates. In this context, temperature modulates the system's behavior by dictating its kinetic pathways.

A particularly insightful class of control parameters involves constraints that drive the system away from equilibrium. The most illustrative case is thermodynamic equilibrium, which—absent any phase transitions—is typically characterized by a single, static state devoid of macroscopic dynamical activity. This equilibrium can serve as a useful reference point. Deviations from it, induced by gradients in temperature, pressure, concentration, or other variables, act as control parameters to the extent that they generate qualitative changes in system behavior.

By examining the interface between a system and its environment, such as the boundary defined by a cell membrane, we can begin to understand how modifications in external constraints (i.e., changes in control parameters) may lead to the emergence of qualitatively new states. The extent to which these transitions produce novel properties represents a central component of the phenomenology of complexity.

3.3 Complexity Evolution

A system's response to variations in a control parameter reveals that increases in complexity typically do not occur in a linear or gradual fashion. Rather, complexity often unfolds as a discontinuous and non-monotonic process, marked by a sequence of transitional and evolutionary shifts between states. These transitions, previously discussed in terms of bifurcations, are frequently captured by universal models known as normal-form bifurcations, and are closely associated with concepts such as instability, chaos, and unpredictability.

Some of these transitions, often referred to as catastrophic events (Arnol'd, 1992), are not directly anticipated by the fundamental laws of physics or their mathematical formulations. As Feynman once noted regarding the Schrödinger equation, the equations themselves are continuous and smooth, offering no hint of the abruptness these phenomena can exhibit. To use a colloquial term aligned with the spirit of this work, as outlined in Chapter One, these events emerge: they are sudden, unexpected, and profoundly surprising.

One illustrative example is the cusp bifurcation, developed by mathematician René Thom and later applied to various domains such as behavioral theory, sociology, and linguistics (Arnol'd, 1992). This model offers a qualitative framework for understanding the dynamics described above. In this scenario, for each value of a control parameter *p*, the system displays a particular state, represented by a variable *u*—which could correspond, for instance, to chemical concentration, population density, or another macroscopic measure. When p lies below (or above) a critical threshold p^* (minimum or maximum), the system exhibits only one possible state. This state is not only singular but also highly stable, in the sense that it can absorb or regulate both external disturbances and internal fluctuations, the latter arising from ongoing interactions among its fundamental components. These influences, external and internal alike, are inescapable for any natural system.

Under such stable conditions, the notion of complexity, at least in the Gell-Mann sense, or in closely related formulations, has limited relevance. The system does not exhibit change or surprise; it remains in a stationary or quasi-stationary state, and its dynamics are relatively simple to model.

However, this picture changes dramatically once the control parameter exceeds the critical threshold p^*. Beyond this point, the previously stable and unique state becomes unstable. When subjected to perturbations, the system begins to act as an amplifier: it leaves its initial reference state and transitions toward one or more new behaviors. While these behaviors may converge back to $p = p^*$ at the bifurcation point, they diverge as p moves further beyond the critical threshold. This is the hallmark of a bifurcation, a phenomenon made possible only through the non-linearity of the system's underlying laws of evolution, which in turn allow for the coexistence of multiple solutions beyond p^*.

To understand its catastrophic nature, as previously envisaged, it is important to take into account the following elements:

- Finite Precision in Observation: Every experimental measurement is inherently subject to finite precision. Observing the state of a system for a given value of a contro*l parameter does not yield a single, exact value on the axis of that parameter. Instead, the measurement spans a small interval—a region of uncertainty—characterized by a probability distribution with a mean value and variance. The system, when considered across all relevant variables, lies somewhere within this uncertain interval, though its precise position remains indeterminate.
- Indeterminacy Beyond Criticality: Near and beyond the critical threshold p^*, a range of possible states becomes accessible to the system. Crucially, no information available to the observer allows for predicting which state the system will ultimately adopt. Chance events, such as evolutionary accidents, as discussed earlier, and internal fluctuations guide the system's trajectory. In this context, the system engages in an implicit exploration of its state space, potentially passing through unstable states, until a fluctuation stabilizes one of them. This stabilization marks a historical bifurcation: any subsequent evolution will be constrained by this critical event. This process, closely linked to the concept of emergence, has a temporal dimension, emergent properties unfold over time from pre-existing

configurations. The appearance of distinct living organisms through evolutionary and self-organizing processes offers a compelling example.

- Sensitivity and Loss of Predictability: For the observer, the extreme sensitivity of the system to parameter values underscores the limits of prediction. Systems that appear identical, within the bounds of measurement uncertainty, can evolve into dramatically different final states. This sensitivity serves as a hallmark of complexity and unpredictability.

Together, these elements define the foundational mechanism for the generation of complexity. In most real-world cases, this mechanism is only the first stage in a sequence of successive bifurcations that significantly increase the number of potential system behaviors. Over time and often across space, the system's properties evolve in ways that become increasingly difficult to describe through algorithmic or computational models.

As this descriptive difficulty increases, so does the amount of information required to capture the system's behavior, contributing to what is often referred to as its Effective Complexity. Importantly, this growth in complexity does not necessarily arise from random external noise or uncorrelated internal fluctuations. Instead, it may emerge from deterministic laws that are intrinsic to the system, as seen in classical spatiotemporal chaos.

One illustrative example is the case of two identical systems, such as double pendulums, that differ only by a minuscule variation in initial conditions—for instance, a 0.1 degree difference in starting position. Despite this seemingly trivial discrepancy, their trajectories soon diverge dramatically. This behavior illustrates a defining feature of complex systems: sensitivity to initial conditions. From an epistemological standpoint, both systems would be treated as indistinguishable at the outset, since the difference falls below the threshold of any practical measurement. Yet as time progresses, the divergence between their states grows, often exponentially, even when averaged across a range of initial conditions. Once this divergence exceeds the resolution of the measuring apparatus, the systems become effectively distinct in the eyes of the observer.

As a result, it becomes impossible to predict the long-term evolution of the system beyond a certain horizon. This unpredictability, which stems from the amplification of microscopic uncertainties, is a hallmark of chaotic dynamics.

This gives us a second important reason to reconsider the limits of prediction in complex systems. All the available evidence, especially from nonlinear dynamics and chaos theory, suggests that it is not possible to anticipate either the number or the nature of the evolutionary paths such systems might follow.

Beyond their limited predictability, complex systems challenge the adequacy of universal and reductionist explanatory frameworks. Yet, paradoxically, it is precisely these limitations that make complex systems so intellectually compelling. Despite the challenges they present, the methodologies developed to study complexity, along with their inherently interdisciplinary perspective, allow us to make sense of a broad array of phenomena. It is within these apparent limitations that much of the beauty of complex systems resides.

3.4 Control Parameters

The macroscopic level of description is usually the most familiar in which complexity manifests, either through the generation of different states (solutions, which are the results of models employing mathematical equations in the field of complexity science) that show unexpected structural and dynamical properties, as is the case of classical chaos. As discussed above, bifurcation processes are one of the basic elements of complexity, since it provides the fundamental mechanism for the emergence of new possibilities from regimes corresponding to very simple behavior of the system. To have epistemic access to its structure, that is, to the type of bifurcation that the system presents in particular in a given region of parameters, we need to solve the evolution equations.

Regarding exact solutions, it is notable that this task is usually beyond the scope of the techniques currently available in science, except in certain particular cases. Therefore, we need to resort to qualitative arguments and approximations, which, as the attentive reader will imagine, immediate impact on measures like the Effective Complexity. We see here then, a whole universe of possibilities for the occurrence of weak emergent properties. Note that weak emergence is compatible with the deductibility of the phenomenon from its bases, although higher-level patterns may be unexpected (Chalmers, 2006a, 2006b). It is interesting at this point to quote here the vision of Mark Bedau, (Bedau, 1997; Bedau & Humphreys, 2008), who has argued that the characteristic of the weak emergence is that although the microscopic phenomena of complex systems are, in principle, ontologically and causally reducible to microscopical phenomena, their reducible explanation is intrinsically extremely arduous, the final states of the system are only available by indirect derivation, through the simulation of the micro-dynamics of the system and external conditions.

Moreover, it is important to stress out that in many cases the evolution equations of some systems do not refer to fundamental dynamics of the basal elements that constitute it (usually called as first principle dynamics), which touches another important point in the discussion of the characteristics at higher levels, given that they do not only refer to the impossibility of understanding of solutions given the inability to find them analytically but also that the understanding of these systems is only approached through the analysis of phenomenological equations. An excellent example of the latter is the case of granular systems. In a paper from the beginning of this century, Aranson and Tsimring dedicate a complete section to the study of "Phenomenological Models" which can be used to understand the Effective Complexity of these systems (Aranson & Tsimring, 2006). Notice that I am referring to Effective Complexity for the case of the best theory (phenomenological) to understand these systems.

Returning to the previous context, the bifurcation analysis is a very useful and powerful technique that allows building the solutions in a perturbative way, near the bifurcation point. Given the importance of this in the present study, making the connection between complexity and emergence, it is worth discussing this topic with some detail.

The mentioned procedure, described here for the study of bifurcations where the reference state is a steady-state solution, can be summarized as follows.

Suppose we have a set of evolution equations of a relevant variable to study. By a standard method known as linear stability analysis, we can determine the values of the parameters for which a certain reference state changes from asymptotically stable to unstable. Stability is essentially determined by the response of the system to disturbances or fluctuations acting on a reference state. Therefore, it is natural to write the dynamic laws, of the equations of evolution, in a form in which these disturbances appear explicitly. We can then express the solution of an equation for a certain variable *u*, close to a stationary point, *r*.

The trick of substituting this type of solution in the equation of the general evolution of the system allows that the mathematical description (the theory) of the evolution, close to *r*, will drastically diminish the Effective Complexity of this theoretical description. In this way, a linear description of the evolution of the system is obtained with the first term of this perturbative expansion, and then other terms of higher orders add the non-linearity characteristic near the critical point. Substituting then the equation of *u(t)* in the equations of evolution we can then write for the departure of the systems from the reference state

$$\frac{de(t)}{dt} = f\left(r + e(t), p\right) - f(r, p).$$

To obtain a more transparent form, we expand this equation around the solution *r* and write explicitly the part of this result that is linear, plus a non-linear correction whose structure does not need to be specified at this stage

$$\frac{de(t)}{dt} = Je(t) + h\left(e(t), p\right).$$

Here, *J* is the coefficient of the linear part and h the non-linear contributions. For the typical n multidimensional case, J is replaced by the set of information that defines a linear operator, a square matrix of dimension n, known as the Jacobian matrix. The characteristics of this simple operator indicate the behavior of the system in the vicinity of the reference state.

Note how, using these totally basic tools of calculation, it is possible to write the evolution equation of systems, in principle, extremely complicated, many of them without an apparent analytical solution, of the following form

$$\frac{de(t)}{dt} = Je(t).$$

Thus, in systems that can be described using this approach, the time evolution of the variable near points of equilibrium and bifurcation will be equal to a linear part plus non-linear terms that in principle can be disregarded. It is worth to note that we have reduced the Effective Complexity remarkably. The stability or instability of the system is reduced to a trivial linear problem that is soluble by elementary calculation methods. It is only in the limit case where $u = r$ is stable, but not asymptotically stable, in which the linear approximation could be inadmissible.

The significance of this discussion lies in the remarkable technical reduction of the problem and the corresponding decrease in the amount of information required to explain it. This epistemic operation does not increase the system's Effective Complexity; rather, it leads to a substantial reduction in complexity, particularly in the vicinity of the bifurcation point.

3.5 Universality

A crucial point to emphasize is that bifurcation equations, those describing the system near the bifurcation point, consistently take the same form, regardless of the specific structure of the system's global dynamical laws. In other words, any dynamic system that meets the appropriate conditions can be expressed in a universal form when analyzed near the bifurcation point. As noted in the previous chapter, these equations are referred to as "normal forms" (Guckenheimer & Holmes, 2013). In this sense, normal forms represent a new level of description that is independent of the fundamental evolution equations governing a particular system.

For systems that exhibit dissipative structures, meaning coherent self-organized patterns arising in nonequilibrium conditions, a concept introduced by Nobel laureate Ilya Prigogine, the reduction of full system dynamics to a small set of order parameters is often associated with the work of Lev Davidovich Landau and his theory of phase transitions (Landau, 1937). This reduction also implies a decrease in the system's Effective Complexity.

Given the preceding discussion, and considering that order parameters are the quantities that "perceive" the transition to new forms of complexity, we can conclude that they offer, in principle, a first quantitative measure of weak emergence. As an illustrative case, for a one-dimensional system near a pitchfork bifurcation, the normal form describing the system is as follows:

$$\frac{\partial u(x,t)}{\partial t} = \left(p - p^*\right)u - au^3 + Du\frac{\partial^2 u(x,t)}{\partial x^2}.$$

Where u represents the variable that we want to study, a and D are parameters that characterize the reduction of the complete dynamics to the normal form, coming from the initial evolution equations (or a combination of the original parameters), p is the control parameter, and p^* is the critical point (the critical value of the control parameter). Here the (weak) emergence corresponds to the existence of solutions that appear, and that are non-existent before the possible transition, after p passes through p^* from left to right.

Everything discussed here can be applied to phenomena of symmetry breaking, such as the Rayleigh-Benard cells presented in various current philosophical contributions on emergence (Bishop, 2008; Bedau & Humphreys, 2008). Note that in this type of systems, where the spatial variable is relevant, the previous equation has to be modified to take into account the presence of a so-called phase variable (for a possible rotation of parts of the systems), specifying the different symmetries in which these cells occur in the space.

The general structure of this equation is preserved, but u naturally now becomes an imaginary variable, and in addition, the cubic term is replaced by: $u|u|^2$, where the quadratic term represents the sum of squares of the real and imaginary part of u. Another example in which there is an imaginary variable (a convenient and compact way of to write down the presence of two connected order parameters through its real and imaginary parts) is the bifurcation of periodic solutions over time, in systems with spatial extension. The corresponding normal form equation is

$$\frac{\partial u(x,t)}{\partial t} = (p - p^*)u - (a_1 + ia_2)|u|^2 u + (D_1 + iD_2)u \frac{\partial^2 u(x,t)}{\partial x^2},$$

which differs from the previous one by the explicit presence of imaginary coefficients.

It is also possible to foresee more complex situations in which there are interactions *between* solutions of more complicated bifurcations. These will generate phenomena of secondary, tertiary or even higher order bifurcation. Once again, note how the reduction of the Effective Complexity of systems that can present emergence is reduced significantly thanks to the use of these methodologies.

Finally, in case of dealing with multiple bifurcations, the previous approach can be expanded. However, as is to be imagined, the explicit construction of a normal form for these cases becomes much more complicated and its universality can no longer be guaranteed.

3.6 Basin of Attraction

As previously discussed, the emergence of new features in cases of possible weak emergent behavior can, as a first approximation, be understood through the framework of bifurcation theory. Of course, other approaches to understanding this type of emergence are also possible (Bedau & Humphreys, 2008). However, to better comprehend the dynamics underlying these processes—which may lead to the transformation of existing solutions and the appearance of new behaviors—it is helpful to introduce the concept of an attractor. In systems influenced by fluctuations, this is sometimes referred to as a non-equilibrium potential (Fuentes & Miguel, 2013, 2016; Fuentes, 2016).

Closely related is the concept of a basin of attraction, which is particularly useful in understanding how a system behaves dynamically within a certain region of its parameter space. Using formal mathematical modeling—such as the control parameters and bifurcation mechanisms discussed earlier—we can begin to interpret the emergence of new features as inevitable outcomes of the basal interactions between constituent elements. The stability of these features and the possible pathways toward new solutions can be analyzed by assigning probabilistic weights to the different options available to the system.

To illustrate the usefulness of this concept, a basin of attraction can be thought of as a surface or landscape on which the system evolves. The system tends to "move" along this landscape and eventually settles into the attractor—a region where it remains stable, even in the presence of internal or external disturbances.

In mathematical terms, an attractor is the set of values toward which the system's time-dependent behavior converges. This convergence occurs from a broader surrounding region, known as the basin of attraction. Any system state sufficiently close to the attractor will remain near it over time, which also reflects the system's robustness.

For attractors in finite-dimensional systems, the relevant evolving variables are typically expressed as an n-dimensional data vector. In such cases, the attractor is a subset of n-dimensional space. When the evolving variables are two-dimensional or three-dimensional, the attractor can often be represented geometrically.

Attractors can take various forms: they may consist of a single point, a finite set of points, a continuous curve, or even a more complex structure, such as a fractal known as a strange attractor. In the simplest case, when the evolving variable is scalar, the attractor is a subset of the real line.

From an epistemological perspective, the significance of this concept becomes clear when considering that the description of attractors has been one of the central achievements of chaos theory. Even in the face of long-term unpredictability, if the system satisfies certain conditions—such as convergence criteria—it is still possible to determine that the system will remain confined within a specific region of n-dimensional space defined by the strange attractor.

The trajectory of a dynamical system within its attractor is not subject to any specific constraints beyond the requirement that it stays within the attractor. These trajectories may be monotonic, periodic, or chaotic. Closely associated with the concept of an attractor are the ideas of invariant sets and limit sets, which further characterize the long-term behavior of complex systems:

- An invariant set is a set that evolves in *itself* under the underlying dynamics. That is, if the element w of set C evolves, the final state of this state will be, say, q, where q is another element of set C. Note that attractors can contain invariant sets.

- A limit set is a set of points such that there exists some initial state that ends arbitrarily close to an established limit, either by a region or a collection of points, when time goes (or tends) to infinity.

Attractors are a particular type of limit set, but not all limit sets qualify as attractors. It is possible for certain points in a system to converge toward a limit set, while other nearby points, if slightly perturbed, may not return to that region. This distinction highlights the need to carefully characterize the stability of a system's long-term behavior.

A clear example of this can be found in the case of a damped rigid pendulum, which features two invariant points. The first, labeled m, corresponds to the position of minimum height. The second, labeled q, corresponds to the maximum height. The point m is a limit set because system trajectories tend to converge toward it. In contrast, point q is not a limit set and is also not an attractor, even though it is stationary and unstable.

Due to energy dissipation from air resistance, the point m also functions as an attractor. If dissipation were absent, m would no longer serve this role. Some authors argue that an attractor must have positive measure, which would exclude a single point from qualifying as an attractor. However, this technical debate is not central to the present work.

What is especially important for our purposes is to observe how emergent properties may arise with differing degrees of robustness. Some features remain stable in the presence of fluctuations, structural changes within the system, boundary constraints, or variations in initial conditions. Others may prove more fragile. Understanding this contrast is essential for studying how complexity evolves and persists in dynamic environments.

3.7 Type of Attractors

From a physical perspective, attractors are regions or subsets of the phase space of a dynamical system. Until the 1960s, it was generally believed that attractors were simple geometric subsets of this space. These could take the form of points, lines, surfaces, or regular regions within three or more dimensions.

However, attractors can exhibit a wide variety of geometric configurations within phase space. When these sets, or in physical terms the trajectories of particles, cannot be easily described as straightforward combinations of basic geometric forms such as lines, surfaces, or spheres, the attractor is referred to as strange. This term reflects the complex and often fractal structure of such sets, which defy classical geometric intuition.

3.7.1 Fixed Points

An excellent example of this type of point is the final position reached by a classical damped pendulum.

From a more technical perspective, a fixed point of a function F or a transformation is a point for which the application of F returns the same value. In other words, the point remains unchanged under the transformation. If we consider the evolution of a dynamical system as a sequence of transformations, then there may or may not exist a point that remains fixed throughout each transformation. The final state toward which a dynamical system evolves corresponds to a fixed point of attraction of the system's evolution equation.

However, as is easy to observe, the fixed point of a given dynamical system is not necessarily an attractor. This can be clearly seen in the case of a rigid pendulum with friction. In this system, there are two fixed points: the point m, representing the position of minimum height, and the point q, corresponding to the position of maximum height. Only one of them functions as an attractor.

It is worth noting that even a simple physical phenomenon such as classical friction continues to be an active field of study. It also serves as an excellent example of the multiple perspectives and levels of coarse graining that exist in physics, ranging from quantum interactions to macroscopically emergent laws (Marino et al., 2017).

3.7.2 Attractors of Finite and Infinite Number of Points

In a discrete-time system, an attractor can take the form of a finite (or infinite) number of points that are visited in temporal sequence. Each of these points is called a periodic point. This example can be illustrated by the so-called logistic map that, depending on its specific parameter value, can have an attractor that consists of $2n$ points, $3 \times 2n$ points, etc., for any value of n. Its equation is the following.

$$x_{n+1} = rx_n (1 - x_n)$$

This case is very useful, due to the explicit connection with a control parameter, and the relation of complexity and emergence (see below). The extensive conceptual discussion around this "simple" mathematical model, is remarkable (Mäki et al., 2015).

3.8 Variation of the Control Parameter and the Complexity Measure

As we have seen in the previous sections, we can understand systems studied by means of mathematical expressions using the concept of dynamic systems, attractors and control parameters. The latter will play a very important role in what will be discussed next, which will have a direct impact on the emergence concept proposed for these systems.

Suppose that we want to give some idea of the complexity of a given system. Suppose we have the time series "generated" by the system, say n number of points, and we have this for two different cases that are related with two different values of some (control) parameter of the system. Let us think that the first data set corresponds to the periodic region. The time series is of the type:

{... 0.826941, 0.500884, 0.874997, 0.38282, 0.826941, 0.500884, 0.874997, 0.38282, 0.826941, 0.500884, 0.874997, 0.38282, 0.826941, 0.500884, 0.874997, 0.38282, 0.826941, 0.500884, 0.874997, 0.38282, 0.826941, 0.500884, .. .}

The system is evolving in a periodic zone, that is to say, during the iterations the series repeats itself. Now, if we change the control parameter of the system, so that it is in a chaotic region, as mentioned in the preceding sections. The time series could be something like:

{... 0.575668, 0.977097, 0.0895126, 0.326, 0.878896, 0.42575, 0.977948, 0.086263, 0.315287, 0.863524, 0.471401, 0.996728, 0.0130434, 0.051493, 0.195366, 0.628792, 0.933651, 0.247789, 0.745558, 0.758805, 0.73208, ...}

We can then ask ourselves what would be the complexity of the system in relation to these new data. If, as we have assumed, we only know the time series and we know nothing about the mechanism that generates them, we can then make a model that generates approximately these two times series: we will try to make a minimum program that generates both series.

The following question (which, as I believe it happens many times in the scientific agenda, depends on the observer) immediately arises: are all the terms of the series important? how we can choose the noise terms when describing the phenomenon, and what will be the coarse-graining?

To motivate the discussion, let us consider only the first case. In this situation, we would describe the time series as periodic, and we might be inclined to treat all data points as equally significant. Based on that assumption, it becomes relatively straightforward to reduce our explanation to a mathematical model of the following type:

Repeat the cycle: {0.826941, 0.500884, 0.874997, 0.38282}

While in the second case, which involves a chaotic time series, we could interpret the situation in one of two ways:

- The time series appears noisy, with all of its points behaving randomly. In this case, its Effective Complexity would be considered null.
- Alternatively, we might assume that every point in the time series is significant and should be modeled individually. Under this interpretation, the simplest program or model capable of reproducing the data would be the time series itself. Consequently, the Effective Complexity would be extremely high, especially for very long sequences generated by the system.

This leads to an epistemological challenge that centers on the interpretation and intention of the observer when modeling a dataset, particularly when the underlying system is unknown. This assumes a one-to-one correspondence between system and model, although that issue will not be addressed here. In such situations, it becomes tempting to classify the observed pattern—whether chaotic or periodic—as an emergent property of the system, depending on the perspective of the observer (Fuentes, 2014).

In the following section, drawing on all the considerations discussed so far, I will present my proposal as a possible definition of emergent properties in any system that allows for a meaningful description, whether mathematical or otherwise.

As a final note, both time series referenced above are generated by the simple logistic map, using different values of the parameter r.

3.9 Complexity Change and Emergent Properties

As discussed so far, we can see that the control parameter is closely tied to the theoretical framework used to explain the phenomena. As previously mentioned, examples of control parameters include the temperature difference in the Rayleigh-Bénard convection system, the diffusion ratio in Turing's pattern formation, mass in gravitational systems, and the number of interconnected cells in a neural network, among others.

Before addressing the relationship between complexity and emergence, we need to introduce a preliminary concept: Parametric Effective Complexity, abbreviated as P-Effective Complexity.

P-Effective Complexity, denoted as Cp, is a function that assigns a value of Effective Complexity to each specific value of the control parameter p within a given model that describes the system under study.

As noted earlier, and applicable in this case as well, P-Effective Complexity is determined by identifying the regularities in the signal present in the data while excluding noise. The distinction between signal and noise is made such that the extracted signal represents statistically significant features that are typical outputs of the system, while the noise lacks such statistical significance.

In the definition of Cp, the control parameter becomes an explicit variable. This step is essential, as it introduces critical implications for the concept of emergence that will be addressed next.

With these foundational ideas in place, we can now define what is meant by an emergent property in systems that can be modeled using dynamical equations, algorithms, or similar frameworks, provided that at least one control parameter is relevant to the appearance of the property in question. It is important to recall the universal features that justify this approach, such as nonlinearity, bifurcation theory, and attractor dynamics, which support the applicability of this notion.

Parametric emergent property, denoted as Ep, refers to a property that is considered emergent at a critical value of the control parameter p^*, if the corresponding function Cp exhibits a step-like discontinuity. In other words, if the complexity on one side of the critical parameter value is significantly lower than on the other side, the property is identified as emergent.

This leads to an important epistemological consequence. In cases where empirical observations do not align with theoretical predictions, we are confronted with a theoretical anomaly. Such discrepancies indicate that either the model must be revised or the framework reinterpreted to account for the observed emergent behavior. It is important to note that by making the control parameter explicit, we are also identifying what the observer considers to be relevant. The definitions of Cp and Ep are meaningful in this context, but their use may become unnecessary once there is consensus regarding what is being referred to, at which point it would be sufficient to refer only to Effective Complexity.

References

Agazzi, E., & Montecucco, L. (Eds.). (2002). Complexity and emergence: Proceedings of the annual meeting of the international academy of the philosophy of science, Bergamo, Italy, 9–13 May 2001. World Scientific.

Aranson, I. S., & Tsimring, L. S. (2006). Patterns and collective behavior in granular media: Theoretical concepts. *Reviews of Modern Physics, 78*(2), 641–692.

Arnold, V. (1992), Catastrophe Theory, Springer Berlin, Heidelberg, https://doi.org/10.1007/978-3-642-58124-3

Bedau, M. A. (1997). Weak emergence. *Noûs, 31*(s11), 375–399.

Bedau, M. A., & Humphreys, P. E. (2008). *Emergence: Contemporary readings in philosophy and science*. MIT Press.

Bishop, R. C. (2008). Downward causation in fluid convection. *Synthese, 160*(2), 229–248.

Campbell, D. T. (1974). Downward causation in hierarchically organised biological systems. In *Studies in the philosophy of biology* (pp. 179–186). Macmillan Education.

Chalmers, D. J. (2006a). *The re-emergence of emergence. Chapter strong and weak emergence*. Oxford University Press.

Chalmers, D. J. (2006b). *Strong and weak emergence. The reemergence of emergence* (pp. 244–256). Oxford University Press.

Fuentes, M. A. (2014). Complexity and the emergence of physical properties. *Entropy, 16*(8), 4489–4496.

Fuentes, M. (2016). Attraction basins and unlimited semiosis. *Journal of Humanities of Valparaiso, 7*, 77 86.

Fuentes, M. A., & Miguel, H. (2013). Self-generated dynamic landscape: The message-receiver interaction case. *Physica A: Statistical Mechanics and its Applications, 392*(10), 2492–2497.

Fuentes, M., & Miguel, H. (2016). Dynamic landscapes to model communication and learning. *Scientiae Studia, 14*(1), 65–94.

Gallagher, S. (2012). What is phenomenology? In *Phenomenology* (pp. 7–18). Palgrave Macmillan.

Godfrey-Smith, P. (2000). Niche construction in biological and philosophical theories. *Behavioral and Brain Sciences, 23*(1), 153–154.

Guckenheimer, J., & Holmes, P. J. (2013). *Nonlinear oscillations, dynamical systems, and bifurcations of vector fields* (Vol. 42). Springer Science & Business Media.

Ingólfsson, H. I., Lopez, C. A., Uusitalo, J. J., de Jong, D. H., Gopal, S. M., Periole, X., & Marrink, S. J. (2014). The power of coarse graining in biomolecular simulations. *Wiley Interdisciplinary Reviews: Computational Molecular Science, 4*(3), 225–248.

Kilicay, N. H., & Dagli, C. H. (2003). Emergence and artificial life. In engineering management conference, 2003. IEMC'03. Managing technologically driven organizations: The human side of innovation and change (pp. 580–584).

Laland, K., Uller, T., Feldman, M., Sterelny, K., Müller, G. B., Moczek, A., et al. (2014). Does evolutionary theory need a rethink? *Nature, 514*(7521), 161–164.

Landau, L. D. (1937). On the theory of phase transitions. *Zhurnal Eksperimental'noi i Teoreticheskoi Fiziki, 7*(19–32).

Levitt, M., & Warshel, A. (1975). Computer simulation of protein folding. *Nature, 253*(5494), 694–698.

Marino, J., Recati, A., & Carusotto, I. (2017). Casimir forces and quantum friction from Ginzburg radiation in atomic Bose-Einstein condensates. *Physical Review Letters, 118*(4), 045301.

Mäki, U., Votsis, I., Ruphy, S., & Schurz, G. (Eds.). (2015). *Recent developments in the philosophy of science: EPSA13 Helsinki*. Springer.

Ohno, S. (1973). Ancient linkage groups and frozen accidents. *Nature, 244*, 259–262.

Schöll, E., & Schuster, H. G. (Eds.). (2008). *Handbook of chaos control*. Wiley.

Chapter 4
Evolution of Scientific Theories and Predictive Epistemology

4.1 Introduction

This section examines how scientific theories evolve in response to growing empirical information, and how tools from complexity theory, can offer a new perspective on theory choice. Rather than recounting traditional debates on scientific revolutions, we focus on how emergent patterns and discontinuities in complexity may signal epistemological thresholds, offering a foundation for a predictive epistemology that anticipates theoretical change.

4.2 Around Copernicus

As is well known, the Copernican revolution was made possible through the contributions of many influential predecessors, in addition to the work of the astronomer whose name it bears. It could be argued that Copernicus himself was not the most significant contributor to our current understanding of the universe. Many others played crucial roles in shaping this view (DeWitt, 2011). Nevertheless, the work of Nicolaus Copernicus marked a pivotal shift in cosmological thought, enabling a new way of perceiving the universe and the human experience within it. This shift laid the groundwork for the theories that modern scientists continue to explore and refine.

It is worth noting that the roots of the Copernican revolution may in fact reach back as far as the sixth century BCE. The theory presented in De Revolutionibus Orbium Coelestium was influenced by classical thinkers such as Heraclides Ponticus, Plato, Pythagoras, Aristotle, and Ptolemy (Dreyer, 1953). Over time, these ideas evolved further through the contributions of numerous astronomers and

M. Fuentes, *Complexity, Emergence and the Evolution of Scientific Theories: Towards a Predictive Epistemology*, SpringerBriefs in Philosophy,
https://doi.org/10.1007/978-3-031-99994-9_4

physicists, including Novara, Tycho Brahe, Kepler, Galileo Galilei, Newton, and even Einstein, each of whom played a role—albeit to varying and difficult-to-quantify degrees—in the ongoing development of our cosmological models.

4.3 Revolutionary Changes

The Copernican revolution is closely associated with the collapse of long-standing scientific and cultural frameworks that had shaped the collective mindset of earlier times. It marked the end of a conceptual tradition that had endured for millennia. As is well known, Kuhn (1957) chose this particular shift as a central example in his analysis of how scientific paradigms evolve, emphasizing that astronomical concepts follow similar patterns of change as those in other scientific disciplines. However, few scientific revolutions have had such a profound impact on non-scientific thought as the Copernican revolution. Comparable cases include Darwin's theory of evolution, Einstein's theory of relativity, and Freud's contributions to psychology. This influence is understandable, since astronomy has always been deeply intertwined with cosmological and existential reflection.

Before Copernicus, the theories that sought to explain the world and the evolution of observable phenomena—such as eclipses, planetary positions, and the seasonal cycle—were regarded with a level of trust comparable to how we perceive scientific theories today. Although these pre-Copernican theories were often complex in their attempts to account for detailed observations, the underlying principles that supported them were generally aligned with prevailing beliefs and religious views. In that sense, they offered a kind of conceptual simplicity that was effective in providing explanations for a variety of natural phenomena, including those as seemingly straightforward as the fixed stars or the alternation between day and night.

It is important to recognize that many distinct cosmologies existed for thousands of years. Cultural practices and beliefs related to celestial observations were deeply rooted in civilizations such as the Egyptians, Mesopotamians, and Babylonians. In Babylonian cosmology, for example, the Earth and the heavens were conceived as part of a unified spatial whole, often imagined as circular or spherical, with references to the circumference of both heaven and Earth. Contrary to later models, Babylonian thought did not center around a geocentric view. Instead, their cosmology suggested that the universe revolved around a central point (Hetherington, 1993). The Sumerians, who preceded the Babylonians, also envisioned a complex cosmos that included multiple heavens and multiple lands (Hetherington, 1993).

It is possible that the ancient Greek inclination toward symmetry contributed to the formulation of the spherical model of the universe. This model integrated Babylonian astronomical records and the observations of thinkers such as Anaximander (Heath, 2004; Popper, 2012), offering a simplified explanation for both the shape and the movement of the sun. Nevertheless, this system exhibited significant anomalies. One of the most notable issues was the difficulty in explaining the observed motion of the planets. This is reflected in the fact that the term "planet," which remains in use today, derives from the Greek word planētēs, meaning "wanderer" (Carroll & Ostlie, 2017).

A detailed treatment of this topic will be presented later. For now, it is sufficient to note that the retrograde motion of the planets, which was already well known in antiquity, consistently challenged the explanatory power of contemporary models attempting to describe the universe and the position of celestial bodies within it.

4.4 The Geocentric Model

As is well known, this model dates back to the early stages of Greek astronomy and philosophy. It can already be found among the pre-Socratic thinkers (Fraser, 2006). In the sixth century BCE, Anaximander proposed a particularly original cosmology and physical theory. According to Popper (1958), this idea of Anaximander represents one of the boldest, most revolutionary, and most significant ideas in the history of human thought. In his model, the Earth is a finite object shaped like a cylinder that floats freely in space. As explained by Rovelli (2011), it does not fall because there is no preferred direction for it to fall, and it is not subject to the dominance of any other body. The model also describes the Sun, Moon, and stars as revolving around the Earth in circular paths. Immense wheels were introduced to support these celestial bodies, similar to the wheels of a cart. These wheels were envisioned as hollow and filled with fire. They were also perforated on the inner surface, facing the Earth. The light from the Sun, Moon, and stars was thought to be the fire seen through these openings. According to this model, the stars were carried by the innermost wheels, the Moon by a middle wheel, and the Sun by the outermost one.

Pythagoras, on the other hand, proposed that the Earth was spherical based on the observation of eclipses. However, he and his school did not place the Earth at the center of the universe. Instead, they suggested that it revolved around an invisible central fire. Philolaus, an important figure in the Pythagorean tradition, can be seen as a precursor to Copernicus for his proposal that the Earth should not occupy a privileged position in the cosmos. He developed a system named after him, in which the Earth no longer stood at the center, although the model was not heliocentric. Rather, it placed a central fire at the center of the universe. This fire, according to the theory, was invisible and not identified with the Sun. Aristotle recorded the following (Guthrie, 1939): "It remains to speak of the Earth, of its position, of the question of whether it is at rest or in movement, and of its form. As for its position, there is some difference of opinion. Those who consider the entire sky to be finite say it is in the center. But the philosophers known as Pythagoreans take the opposite view. In the center, they say, is fire, and the Earth is one of the stars. It creates night and day because of its circular motion around the center. They also describe another Earth in opposition to ours, which they call the counter-Earth or Antichthon."

It is remarkable that although these models were far from providing accurate explanations for the wide range of observable phenomena, they were already quite sophisticated. They involved a considerable amount of information concerning the positions of celestial objects, their apparent speeds, and their interactions. In these early cosmological models, we find both the phenomena to be explained and the conceptual elements introduced to account for them, such as wheels or central fire.

In the fourth century BCE, the Greek philosophers Plato and his student Aristotle wrote extensively based on the geocentric model. According to Plato, the Earth was spherical and stationary at the center of the universe. The stars and planets were thought to move around it in geometric shapes such as spheres or circles. The order of these celestial bodies, moving outward from the center, was the Moon, the Sun, Venus, Mercury, Mars, Jupiter, Saturn, and finally the fixed stars, located on the celestial sphere.

In the "Myth of Er," a section of The Republic, Plato described the cosmos using metaphors and mythical elements. Eudoxus of Cnidus, who worked alongside Plato, developed a less mythological and more mathematical approach to the movement of celestial bodies. He followed the Platonic principle that all celestial phenomena could be explained through circular and uniform motion.

Adding the contributions of Aristotle, the model became even more structured. In the Aristotelian system, the spherical Earth is positioned at the center of the universe, and the other celestial bodies are embedded in a series of transparent, concentric, and rotating spheres. These crystalline spheres move at different uniform velocities and account for the observed motion of the celestial bodies. Aristotle believed that the Moon resided in the innermost sphere and was thus the closest to the Earth. This explained the lunar phases and the presence of dark areas on the Moon, which he interpreted as physical features.

Aristotle further expanded his cosmology by proposing a system of natural motion based on the elements. The Earth was considered the heaviest element and therefore moved naturally to the center. Water followed, while air and fire moved upward away from the center. Beyond the region of fire were the solid spheres made of a fifth element, the ether, in which the celestial bodies were embedded.

Once again, it is striking to see how rich in conceptual and informational content these ancient cosmological theories were, especially when contrasted with the relatively limited range of accurate predictions they were able to provide.

4.5 The Data and the Adjustment of the Model

As mentioned earlier in general terms, adherence to the geocentric model was largely based on a series of significant observations and on the fact that the explanations offered a kind of intuitive clarity that made sense to both contemporary intellectuals and the general public. One of the central arguments was related to the absence of observable stellar parallax. If the Earth were in motion, one would expect to observe a displacement in the position of the fixed stars over the course of a year. That is, the shapes of constellations should change noticeably as the Earth moved through space. If no such movement was observed, then it would imply that the stars

were vastly more distant than the Sun and the planets, to a degree that rendered their motion undetectable with the instruments of that time.

In fact, the stars are indeed much farther away than Greek astronomers had imagined, making the parallax effect extremely subtle. Stellar parallax was not detected until the nineteenth century. The first successful measurement was carried out by Friedrich Bessel in 1838 at the Königsberg Observatory, using the star 61 Cygni as a reference. This achievement was made possible through a technical innovation known as the Fraunhofer heliometer (Hirshfeld, 2002).

Returning to the discussion, another observational argument in favor of the geocentric model involved the apparent consistency of the brightness of Venus. Its relatively uniform luminosity was interpreted as evidence that Venus remained at a constant distance from Earth. This observation seemed more compatible with a geocentric model, which placed Venus on a circular path around Earth, rather than one in which it orbited the Sun.

A further important argument came from the observation that terrestrial objects tend naturally to fall toward the center of the Earth. Since the Earth was believed to be at the center of the universe, this observation appeared to support the geocentric framework. Taken together, these pieces of evidence provided substantial justification for placing the Earth in a central and privileged position in the cosmos.

Nevertheless, no conceptual system can claim permanent validity. The minor inaccuracies present in the Aristotelian model of concentric spheres, which became more evident over time, gradually laid the foundation for a reevaluation of the geocentric model.

Since the time of Plato, philosophers had reflected on the challenge of integrating the movement of planets, which were perceived as wandering objects, into a universe defined by regularity and mathematical order. The astronomer Ptolemy addressed this issue by studying the peculiar motion of the planets, which at times seemed to reverse direction relative to the fixed stars. This retrograde motion was especially problematic for the existing model. Through detailed observations and theoretical refinements, Ptolemy succeeded in adapting the Aristotelian framework to accommodate these empirical anomalies. His model introduced a system of epicycles and deferents, which allowed for a more accurate representation of planetary trajectories and preserved the general structure of the geocentric worldview.

4.6 Retrograde Motion

As is well known, some planets appear to change their motion relative to the Earth-based observer not only in speed but also in direction. In fact, the planets visible to the naked eye move against the background of the constellations in a way that generally causes them to lag behind the stars. However, during shorter intervals, they appear to move forward relative to the constellations. This phenomenon, known as retrograde motion, had a significant impact on geocentric cosmological theories.

The central challenge, in simplified terms, lies in explaining how such motion could be accounted for from the perspective of a stationary observer using a model based solely on uniform circular motion. In this context, the initial model consists of a set of instructions or rules that function as an algorithm. When this algorithm fails to explain an observable phenomenon such as retrograde motion, it becomes necessary to introduce additional information. This process inevitably increases the model's effective complexity.

At that time, the way to improve the existing model in order to account for retrograde motion and other related phenomena was through the ingenious introduction of coupled circular movements, known as deferents and epicycles.

As will be seen shortly, this addition to the theory proved to be a powerful conceptual tool. By using repeated applications of this method, it became possible to describe any smooth and mathematically continuous curve within the constraints of the system.

4.7 Epicycle and Deferent

The epicycle was a particularly interesting innovation. Derived from the ancient Greek words épi, meaning "upon," and kúklos, meaning "circle," it literally refers to "a circle upon a circle." In the context of celestial dynamics, this term describes a smaller circular path that moves along a larger circular orbit. The epicycle was a geometric model that was used effectively, conceptually speaking, to explain the variations in speed and direction of the Moon, the Sun, and the planets as observed from Earth.

The epicycle constituted a significant addition to the basic geocentric model of celestial spheres. As previously mentioned, it explained the apparent retrograde motion of the five planets that exhibit this phenomenon. It is also important to note that the model accounted for the apparent changes in distance between these planets and the Earth (DeWitt, 2011).

The motion of a celestial body involving a single epicycle consists of two components. The first is the deferent, a large circular orbit centered approximately around the Earth. The second is the epicycle itself, a smaller circular motion carried on top of the deferent. The combination of these two movements produces the complex paths observed from Earth.

It is worth pausing to highlight the hidden power of this technique. When multiple epicycles are added to one another, the resulting trajectories closely resemble those obtained through the mathematical technique now known as Fourier series expansion (Hanson, 1960). In 1822, Joseph Fourier demonstrated that smooth and well-behaved functions can be expressed as an infinite sum of harmonic components (Fourier, 1822).

Returning to the historical development of the epicycle, this addition to the geocentric model was first proposed by Apollonius of Perga toward the end of the third century BCE. It was later developed by Hipparchus of Rhodes in the second century

BCE, and formalized by Ptolemy in his influential astronomical treatise The Almagest, written in the second century of the Common Era (Lankford, 1997).

It is also interesting to note that the principle of epicyclic motion appears in the Antikythera mechanism, an ancient Greek astronomical device designed with a clockwork-like structure. This mechanism was capable of compensating for the elliptical motion of the Moon, which moves faster at perigee and slower at apogee. This compensation was achieved using a system of circular gears housed in a wooden case. The mechanism employed four gears, including a pair coupled in an eccentric manner that approximates Kepler's second law with surprising accuracy (Freeth et al., 2006; Carman, 2011; Carman & Evans, 2014; Anastasiou et al., 2014).

Despite the added explanatory power provided by the use of epicycles, the resulting model increased the Effective Complexity of the original circular geocentric system. Even with this extension, the movements of the planets in terms of both position and velocity were still not predicted with complete accuracy, given the limitations of observational data at the time.

In the Ptolemaic system, the planets were modeled as moving along epicycles mounted on deferents. Both circles rotated in the same direction and were oriented approximately parallel to the plane of the Sun's apparent path, known as the ecliptic. Although the system remained geocentric in its structure, the center of each planet's motion was not located exactly at the Earth. Instead, it was placed slightly off-center at a point called the eccentric.

This adjustment once again increased the Effective Complexity of the geocentric model. However, discrepancies between the model and the observational data persisted. Another refinement introduced by Ptolemy involved the angular velocity of the deferent, which did not appear constant unless it was measured from a specific point. This point, equidistant from the center and the eccentric, was called the equant.

If we examine the evolution of the model over time, we see a clear pattern. Each new observational inconsistency led to the introduction of additional conceptual components. These included deferents, a growing number of epicycles—eventually reaching a total of 80 circular paths according to Palter (1970)—as well as the eccentric and the equant. Without question, the resulting system had a much higher complexity than the earlier, more basic geocentric models.

4.8 The Change to Simplicity

There were, without a doubt, many adjustments and additions made to the geocentric models in an effort to improve their alignment with observational data. These modifications involved the introduction of new conceptual tools, yet they remained within the boundaries of the well-established theoretical framework.

A significant turning point came with the publication of Copernicus's major work, *De Revolutionibus Orbium Coelestium*, or *On the Revolutions of the Celestial Spheres*, which appeared in 1543, the year of his death. This text marked the conclusion of several long-standing astronomical concepts and theories from antiquity, while simultaneously serving as the foundation for a new astronomical paradigm.

In many respects, Copernicus's model did not radically depart from traditional astronomical thinking. The underlying conceptual structure, the mathematical instruments employed, and even the physical assumptions were quite similar to those found in the Ptolemaic system. For instance, the early chapters of the book suggest that Copernicus still held the view that celestial spheres were real. These sections emphasize that the universe is spherical, that the motion of the stars is uniform and circular, and that both the Earth and the seas are spherical in shape.

Despite these similarities, the increasing mismatch between predictions made using the Ptolemaic system and what could be observed with the naked eye led Copernicus to adopt what was then considered a radical position. Regarding the location and motion of the Earth, he broke with Ptolemaic tradition. These departures may be viewed as the principal original contributions of his work. He proposed that the Earth is not stationary, as had long been believed, but instead is in motion. This shift laid the groundwork for a new heliocentric model of the universe and represented a major step forward in the history of astronomy. In his cosmological model Copernicus tell us that the Earth is another planet, revolving around the fixed sun once a year, and on its axis once a day.

The main characteristics of the Copernican theory are:

- The celestial movements are uniform, eternal and circular or composed of several circles (epicycles).
- The center of the universe is close to the Sun.
- Around the Sun, in order of the closest to the most distant, are Mercury, Venus, Earth and Moon, Mars, Jupiter, Saturn, and the fixed stars.
- The Earth has three movements: the daily rotation, the annual revolution and the variation of the annual inclination of its axis.
- The retrograde movement of the planets is explained by the movement of the Earth around the Sun.
- The distance from Earth to the Sun is small compared to the distance to the stars.

De Revolutionibus contains detailed and often laborious mathematical calculations carried out by Copernicus, which provided the foundation upon which a new astronomical tradition could be built. Much of the work's initial success, particularly in the face of opposition from the Church, can be attributed to the fact that it was essentially unreadable to anyone except those with advanced knowledge in astronomy and mathematics. As a result, the theory remained largely unknown for many years outside a very select group of highly educated scholars.

It was not until more than a century and a half later, following the groundbreaking contributions of Tycho Brahe, Johannes Kepler, Galileo Galilei, and Isaac Newton, that the Copernican model gained widespread acceptance among astronomers and scientists.

What becomes evident in this process is a significant simplification of the model and its explanatory mechanisms. With relatively few core concepts, it became possible to account for a broad range of natural phenomena in a more satisfactory and coherent way. It is also essential to highlight the corresponding decrease in the amount of information required to execute the ordered sequence of steps, or algorithm, needed to explain the observed data.

4.9 The Paroxysm of the Geocentric Complexity: The Tycho Brahe's Model

Ironically, Tycho Brahe was a determined opponent of Copernican theory and remained a conservative adherent of Ptolemaic astronomy. His main contributions to science came from the extraordinarily precise observational techniques he developed, which relied solely on naked-eye measurements. He also produced a detailed and systematic record of celestial observations, including the movements of stars and planets.

Brahe proposed an alternative model that combined elements of both the Ptolemaic and Copernican systems. He retained the Ptolemaic idea of placing the Earth near the center, with the Moon and the Sun revolving around it, but adopted the Copernican insight that the other planets orbit the Sun (DeWitt, 2011). This hybrid system was an attempt to preserve the structure of traditional cosmology while incorporating certain empirical advantages of the newer theory. It is an important example of how a model that becomes overloaded with additional information can begin to lose credibility within the scientific community.

In Brahe's system, known as the Tychonic model, Mercury, Venus, Mars, Jupiter, and Saturn revolve around the Sun, while the Sun and the Moon orbit the Earth. Although this system was based on highly accurate observational data, it still required the introduction of epicycles and deferents in order to achieve a level of predictive accuracy comparable to that of the Copernican model (Lankford, 1997). In other words, the geocentric-like model demanded a larger quantity of theoretical adjustments in order to match the explanatory power of a competing theory. In many respects, Brahe's system represented a compromise. It maintained the mathematical strengths of the models presented in *De Revolutionibus* while avoiding the cosmological and theological implications of removing the Earth from its central position in the universe.

Brahe's work provided an extensive set of planetary observations that proved invaluable to later astronomers. Perhaps his most important contribution was that his model created a space in which Copernican theory could be considered by scholars who were not yet ready to accept the radical idea that the Earth is in motion.

His colleague Johannes Kepler, by contrast, was a committed Copernican and a thinker influenced by Neoplatonic philosophy. After many years of sustained effort, Kepler succeeded in building a new planetary model using the precise data collected by Brahe. He used this information not only to address specific problems in *De Revolutionibus* but also to advance one of his most significant arguments: that the

Earth is just one planet among many. This idea fundamentally challenged the traditional notion that the Earth held a unique or privileged position in the cosmos.

4.10 Kepler's Search for Armory

Kepler's first major astronomical work, Mysterium Cosmographicum (The Cosmographic Mystery, 1596), was perhaps the first published defense of the Copernican system. In this book, Kepler observed that regular polygons displayed a ratio between their inscribed and circumscribed circles that seemed to coincide with the ratio of the average distances from the Sun of successive planets. He hypothesized that this relationship might provide the geometric foundation of the universe.

He dedicated considerable effort to finding a unique arrangement of geometric forms that would align with the known astronomical data, including configurations that accounted for the addition of more planets to the system. Kepler began experimenting with polyhedra and discovered that each of the five Platonic solids could be inscribed within and circumscribed around spheres. The radius of these spheres appeared to correspond to the spacing between the orbits of successive planets.

Kepler imagined arranging the five Platonic solids like nested layers, with each solid enclosed within a sphere and enclosing another smaller one. This structure produced six concentric spheres, reminiscent of onion layers, which he associated with the six known planets at the time: Mercury, Venus, Earth, Mars, Jupiter, and Saturn, moving outward from the Sun. By ordering the solids as the octahedron, icosahedron, dodecahedron, tetrahedron, and cube, Kepler found that the spacing of the spheres closely matched the relative distances of the planetary orbits, assuming that the planets orbited the Sun.

He also proposed a formula that related the size of a planet's orbital path to the duration of its revolution. According to his idea, from the inner planets to the outer ones, the increase in the orbital period would be approximately double the difference in orbital radius. However, Kepler ultimately rejected this formula because it did not offer sufficient precision in comparison with the observational data available at the time (Caspar, 1993).

This early cosmological model proposed by Kepler is undoubtedly a remarkable work of intellectual creativity. It touches on many areas of inquiry, including theology, mathematics, and aesthetics. However, as mentioned earlier, it did not achieve the level of precision that Kepler himself expected and that the available data demanded. The empirical record, particularly the accurate measurements provided by Tycho Brahe, had reached a level of quality that could be considered a reliable signal. There was little room left for treating discrepancies as noise or observational error. Consequently, any serious model had to account for all relevant details in the data, which implied an increase in Effective Complexity relative to earlier, purely geocentric models.

Paradoxically, Kepler's effort in this work may be understood as an attempt at conceptual simplification, despite its apparent complexity. His goal was to explain the universe using the ideal harmony of Platonic solids. Yet, the resulting model turned out to be more complex, since its alignment with empirical data was neither direct nor convincing, and the mismatch between theory and observation was increasingly evident.

In 1609, Kepler published *Astronomia Nova*, a work that presented the results of several years of research, with a focus on the motion of Mars. He once again relied on the precise observational data collected by Tycho Brahe.

This book stands as one of the most important works in the history of astronomy. As noted earlier, it provided compelling support for the heliocentric theory and contributed a groundbreaking rethinking of planetary motion. For the first time, Kepler introduced the idea that planets follow elliptical orbits and that their speed varies along the trajectory. These insights marked a decisive step toward modern celestial mechanics and significantly advanced the alignment between theory and observation. *Astronomy Nova* contains two of the Kepler's laws:

1. Planets move in elliptical orbits with the sun in a focus of the ellipse.
2. The speed of the planet changes at each moment such that the time elapsed between two positions is always proportional to the area in the orbit between these positions.

   His famous third law, published in *Harmonices Mundi* (The Harmony of the Worlds, 1619) states that (Schoot, 2001):
3. The relationship between the cube of the length of the semi-major axis of the orbit and the square of the time of its orbital period is the same for all planets.

It is important to note that the first of Kepler's three laws leaves no space for his earlier conjecture regarding the adjustment of planetary orbits through regular solids inscribed and circumscribed by spheres. The precision of the observational data made it clear that the actual trajectories of the planets did not correspond closely enough to the spherical layers associated with the Platonic solids. Therefore, if Kepler or any contemporary astronomer had wished to preserve the vision of a physical universe harmonized by these geometric figures, it would have required the addition of ad hoc computational mechanisms. These would have been needed to reconcile the observed planetary positions with those idealized spherical distances.

Such adjustments would have demanded the inclusion of extra steps to account for the discrepancy between the recorded positions and the predictions of the solid-based model. This would have inevitably led to a substantial increase in the Effective Complexity of the system. The elegance and conceptual appeal of the original idea would have come at the cost of greater informational and computational burden.

By contrast, the new heliocentric model, based on elliptical planetary motion, achieved extraordinary agreement with the available data. It explained irregularities that remained unresolved in the Copernican model and matched Tycho Brahe's measurements with exceptional precision and minimal inconsistency. It is possible to imagine how significant this breakthrough must have felt for Kepler. Abandoning the long-standing assumption of perfect circular motion, he discovered a much simpler and more accurate system. From the perspective of our discussion, this represents a dramatic reduction in Effective Complexity. With only a few fundamental principles, one could now describe the observed data more effectively than ever before.

In addition to the three laws of planetary motion, Kepler proposed a theory of an attractive force that holds the planets in orbit around the Sun. He called this the anima motrix, and described it as a kind of force radiating from the Sun that could attract or repel the planets, thereby explaining the variation in their distances from the Sun. This idea contains conceptual elements that can be seen, in retrospect, as early steps toward the later development of the theory of gravity, which would be formalized by Newton several decades later.

It is also worth mentioning that Kepler's model could not determine the exact position of a planet at any given moment, due to the continuous change in orbital speed. This issue, which came to be known as the "Kepler problem," played an important role in stimulating the development of calculus. The mathematical tools needed to solve this problem were later formulated by Newton and Leibniz.

Another important aspect is the way Kepler treated observational data. He based his entire model on the motion of Mars, and he approached this data with the assumption that it was no longer subject to significant deviations attributable to observational noise. Instead, the data was understood to represent a clean and reliable signal that the theory had to explain. Furthermore, Kepler did not treat this case as a special or isolated phenomenon. Rather than introducing an exception for Mars and increasing the model's complexity, he treated the Mars data as representative of a general rule. The same mathematical model could be applied to all known planets, which in turn contributed to a reduction in the overall complexity of the system. This decision was essential for achieving a unified and elegant planetary theory grounded in empirical data.

4.11 Galileo and the New Worlds

Galileo Galilei also played a fundamental role in the eventual success of the Copernican universe. His discoveries using the telescope introduced new empirical challenges that had to be analyzed within the framework of the emerging heliocentric and elliptical cosmology. Building upon the advances of Kepler and Brahe, Galileo demonstrated that the visible size of the stars was smaller than previously believed. When observed through a telescope, they appeared simply as luminous points rather than extended disks. He also established that the Sun rotates and possesses surface irregularities, much like the Moon.

Among his most significant discoveries were the moons of Jupiter and his detailed observations of Venus, both of which provided strong support for the idea of a moving Earth. If the Earth remained stationary, Venus would always appear in a crescent phase when viewed from Earth. However, Galileo observed that Venus displayed a full set of phases, similar to those of the Moon. This observation led him to conclude that both the Earth and Venus revolve around the Sun. As a result of these findings, and the growing accessibility of the telescope, the Copernican model gained credibility and gradually displaced the Ptolemaic tradition. The arrival of Newton's contributions would later complete this paradigm shift.

Nonetheless, opposition to the Copernican system persisted for many decades. Much of this resistance stemmed from an underlying reluctance to abandon the traditional geocentric worldview. Some of the most prominent and outspoken critics included John Donne and John Milton. Their religious, moral, or aesthetic convictions prevented them from embracing a paradigm that many contemporary scholars had already accepted. They argued that while the telescope might appear to offer new insights, it did not provide definitive proof. This line of opposition continued well into the eighteenth century.

Within the context of our discussion, one of Galileo's most significant contributions was his expansion of the empirical dataset. The unforeseen observation of five celestial bodies orbiting Jupiter posed a serious challenge to existing cosmological theories. The consistent visibility of these moons could no longer be treated as observational noise within the geocentric model. From the standpoint of our framework, Galileo increased the volume of available data and introduced critical information that could not be derived from the algorithms of the prevailing theory.

These observations inevitably led to a substantial increase in the Effective Complexity of the model. The introduction of entirely new celestial systems required theoretical adaptations that significantly exceeded the explanatory capacity of the geocentric framework. Only a radical transformation in the underlying models and theoretical assumptions could restore a lower, and more manageable, level of complexity.

We might say that this escalation in Effective Complexity signaled the onset of a profound conceptual and methodological upheaval that would soon redefine astronomy and the broader scientific worldview.

4.12 Newton's Synthesis

Newton took the next decisive step. He proposed that the motion of celestial bodies, such as the Moon around the Earth and the Earth along with the other planets around the Sun, could be explained by postulating the existence of a force acting on each orbiting body, directed toward the central mass of its orbit. He further hypothesized that this centripetal force decreases with the square of the distance between the objects. Newton then demonstrated that a force with these characteristics naturally results in elliptical orbital motion.

In doing so, he provided a coherent explanation for the irregularities in planetary movement that had remained unresolved. These discrepancies could now be accounted for by his theory of gravitational attraction, which stated that all celestial bodies exert forces upon one another. This theoretical framework not only explained existing anomalies but also contributed, much later, to the discovery of the planet Neptune. The problems first raised by the Copernican revolution thus became the conceptual foundation for the emergence of Newton's universe and directly informed the breakthroughs that followed. To calculate the position and velocity of the planets with precision, one could apply calculus, a mathematical method developed independently by both Leibniz and Newton (Boyer, 1959).

Newton's model was remarkably simple. Only a minimal amount of information was needed to reconstruct a system's past state, explain its present state, or predict its future evolution. The success of this astronomical framework was without precedent in the scientific world of Newton's time. It would take more than two centuries before another theoretical model could surpass its explanatory power and simplicity.

Even today, Newtonian concepts are still widely used due to their computational efficiency. In the vast majority of practical cases, they produce results that are accurate enough for most purposes. Only in scenarios requiring extremely high precision—such as the calculation of Mercury's perihelion or the functionality of satellite-based positioning systems—is it necessary to employ the methods derived from Einstein's theory of relativity, whether special or general, depending on the context. This continued utility highlights just how powerful and enduring Newton's theory remains.

4.13 Perihelion of Mercury

As we discussed earlier, Newton's revolutionary insights into the mechanics of the universe, including his formulation of the inverse square law of gravitation to explain the elliptical orbits of the planets, culminated in a comprehensive vision of the cosmos. This body of work brought closure to the period often referred to as the Copernican revolution. The publication of Principia triggered a fundamental shift in the prevailing worldview, opening new lines of scientific investigation and conceptual inquiry. The scientific community began to perceive the universe once again as a profoundly novel system.

Along with this transformation came the emergence of new phenomena that demanded explanation. These unexpected findings gave rise to new questions and spurred researchers to propose and test potential solutions. However, it is important to emphasize that some issues, such as the appearance of emergent patterns, may originate from the interactions of system components not previously accounted for, or they may reflect properties that are irreducible within the framework of existing theories. In such cases, the challenge leads to the development of new conceptual or computational tools capable of modeling these phenomena more appropriately.

One such challenge arose following Newton's discoveries. A notable discrepancy emerged between the theoretical predictions based on Newtonian mechanics and the observed precession of Mercury's perihelion, as recorded by Le Verrier (Roseveare, 1979). The currently accepted value of Mercury's perihelion precession is 574.6 arcseconds per century, which is approximately 43.6 arcseconds faster than predicted by Newton's theory and slightly faster than the original findings made by Le Verrier.

This anomaly became the focus of intense scientific attention, and for several decades, many of the most respected physicists of the time attempted to resolve it by formulating new theories. Le Verrier initially hypothesized the existence of an undiscovered planet orbiting between Mercury and the Sun. In this approach, the solution involved modifying the initial conditions of the theoretical model by introducing an additional celestial body that would contribute an extra gravitational force, thereby affecting Mercury's trajectory. He named this hypothetical planet Vulcan, and the idea gained support from prominent astronomers such as Babinet, Buys-Ballot, and Kirkwood. Even the Royal Astronomical Society acknowledged the significance of this proposed explanation.

Vulcan seemed to offer a promising resolution to the perihelion anomaly, until new findings came to light. In 1882, Newcomb recalculated the anomaly, determining it to be exactly 43 arcseconds per century. He also identified further inconsistencies in the orbital data for other planets, particularly regarding the nodal positions of Venus. These findings prompted Hall, in 1884, to question the exactness of Newton's inverse square law. He proposed a modification of the gravitational law by adding a correction factor. When he applied this adjusted model to Mercury's motion and calibrated it against observational data, he obtained an exponent of approximately -2.00000016 for the gravitational force, rather than the idealized value of -2. This revised exponent produced more accurate results.

At this stage, the possibility of a more refined law of gravitation was already under serious consideration. The development of new theories in electromagnetism during the late nineteenth century, particularly those advanced by Maxwell and Coulomb, lent further momentum to this line of inquiry. These theories revealed that electric charges interact in ways that depend on their relative velocities, which in turn shape the resulting forces. Building on these insights, Weber, Riemann, Gauss, Clausius, and Ritz proposed various alternative models of gravitational and electromagnetic interaction. Nonetheless, none of these attempts were able to account for a discrepancy greater than 14 arcseconds per century in the perihelion precession of Mercury.

In 1906, von Seeliger proposed a new theory suggesting that the precession of Mercury's perihelion was influenced by the mass of the Sun. This hypothesis received modest support, particularly from De Sitter and Jeffreys, and for a time it appeared to account for the observed anomalies in planetary orbits. However, this explanation was ultimately invalidated following the results of the solar eclipse observed in 1919. It is worth noting that this episode bears resemblance to earlier attempts to preserve outdated cosmological frameworks. Just as Copernicus had once referred to epicycles and deferents as conceptual patches applied to the Aristotelian model of celestial spheres, von Seeliger's theory reflected a similar pattern of augmenting an established theory with additional information when it could no longer explain the available empirical data. These additions served as provisional measures while scientists sought more comprehensive solutions to emerging inconsistencies within their discipline.

Another notable attempt to resolve the anomaly came from Gerber, who approached the problem using a framework reminiscent of Riemann's ideas.

Drawing on an analogy with electromagnetism, Gerber developed a model based on delayed potentials and posited a finite speed for the propagation of gravitational effects, which he identified as equal to the speed of light in a vacuum. Remarkably, this model yielded a correct prediction of Mercury's anomalous precession. However, it failed to gain recognition from the scientific community for many years. This lack of attention can be attributed to several factors. The paper was published in 1898 in a relatively obscure journal, the theoretical foundations of the model were poorly articulated, and subsequent analyses revealed errors in its mathematical formulation. Despite having anticipated some elements of general relativity by nearly two decades, Gerber's work remained on the margins of mainstream scientific discourse. Einstein later acknowledged reading Gerber's article but insisted that it had no influence on the development of his own theory.

Ultimately, it was Einstein's general theory of relativity that superseded Newton's theory of universal gravitation. The new theory not only offered greater accuracy but also succeeded in resolving the specific issue of Mercury's perihelion, along with other longstanding anomalies. Einstein's insight lay in his ability to demonstrate how the geometry of space and time must be curved in order for the laws of gravity to align with observed phenomena. By replacing the flat geometry of Euclid with a dynamic and curved geometric framework, Einstein provided a radically new foundation for understanding gravitation.

This advance represented a profound transformation in the scientific worldview. The inverse square law that had served as a cornerstone of Newtonian mechanics was no longer sufficient in these cosmological contexts. Einstein's theory provided a more robust conceptual and mathematical structure that not only addressed the limitations of previous models but also opened new avenues for future research. In doing so, it permanently altered our understanding of the universe.

4.14 Parametric Complexity and Theory Change

We can observe that, with high probability, during each of the historical periods previously discussed, theories were chosen according to criteria that we can now associate with conceptual and algorithmic frameworks exhibiting the lowest effective complexity of the model or algorithm employed. The complexity of a theory or model must be analyzed by considering the trajectory of the variable that has been designated as the control parameter, which, as previously explained, determines the epistemological discussion regarding the model and its success.

For certain values of this parameter, the complexity of the model remains low, as in the case of classical mechanics outside the relativistic or quantum domains. In such cases, the model becomes the optimal choice when compared to others that display greater complexity. However, for other values of the control parameter, complexity increases sharply, indicating that the algorithm used for making predictions must be supplemented with additional information. Therefore, it is appropriate to consider model complexity as a valuable tool for understanding the processes behind theory selection and the preference for one model over another. This reveals the criteria that may be operating within the scientific community engaged with a particular empirical domain.

The quality of the data set is a crucial factor. Once a data set is established, the best theory is the one that minimizes complexity. It is important to highlight that throughout different historical periods, as the attentive reader will have noticed, there are often competing theories that are equally capable of accounting for the observed data. In cases such as the Copernican theory compared to that of Ptolemy, a possible reduction in complexity leads to a significant simplification, requiring fewer accidental data points, and we see that it was ultimately the theory that gained acceptance. A similar situation occurred in the comparison between Einstein's theory of relativity and Newtonian theory, where numerous attempts by researchers—ranging from the postulation of the hypothetical planet Vulcan to the modification of the exponent in the inverse square law—proved insufficient. In both cases, the adoption of Einstein's theory of relativity resulted in a drastic reduction in complexity, which explains its success and enduring acceptance.

4.15 Towards a Predictive Epistemology

One of the most significant implications of introducing parametric complexity into the analysis of emergent phenomena is the possibility of anticipating theoretical stress before it reaches the level of explicit contradiction. This anticipatory perspective, what in this context we propose to call a predictive epistemology, seeks to establish a framework for identifying when and where a scientific theory is approaching the limits of its explanatory power.

Traditionally, epistemology has focused on the descriptive adequacy of theories and their capacity to reconstruct empirical observations. In contrast, predictive epistemology is oriented toward the diagnosis of failure points, where internal increases in complexity signal a pending need for conceptual or algorithmic revision. In this view, theories are not only judged by their capacity to explain what is already known, but also by how they cope with incoming complexity.

Within this framework, the parametric complexity function plays a central role. As shown in this chapter, we can assign to each value of a control parameter the corresponding effective complexity of the system's regularities. A sharp increase, a staggered discontinuity, in this function suggests the emergence of new properties that are not smoothly assimilated by the existing theoretical framework. This signals the possible presence of a parametric emergent property, but more importantly, it highlights an epistemic tipping point: a moment when the theory may require extension, redefinition, or even replacement.

In practice, predictive epistemology offers a set of tools to detect such transitions. Rather than waiting for anomalies to accumulate post hoc, we can trace forward-looking indicators embedded in the dynamics of complexity itself. In this sense, the theory becomes not only a lens for retrospective analysis but a compass for theoretical evolution.

This approach has both philosophical and practical consequences. It challenges the view that emergence is only legible in hindsight, and instead posits that certain emergent properties can be flagged in advance through complexity metrics. Moreover, it situates emergence not only as an ontological feature of systems but as a dynamic force within the epistemic evolution of science itself.

In the broader context of this work, predictive epistemology serves as a conceptual bridge between the formal treatment of emergence and the historical dynamics of theory change. It invites us to consider emergence not merely as something to be explained, but as something that explains the limits of explanation, and therefore, as a principle for scientific transformation.

References

Anastasiou, M., Seiradakis, J. H., Carman, C. C., & Efstathiou, K. (2014). The Antikythera mechanism: The construction of the Metonic pointer and the back plate spirals. *Journal for the History of Astronomy, 45*(4), 418–441.

Boyer, C. B. (1959). *The history of the calculus and its conceptual development*. Dover Publications.

Carman, C. C. (2011). El mecanismo de Anticitera. Una computadora astronómica de la antigüedad. *Ciencia hoy: Asociación Ciencia Hoy, 21*(123), 32–38.

Carman, C. C., & Evans, J. (2014). On the epoch of the Antikythera mechanism and its eclipse predictor. *Archive for History of Exact Sciences, 68*(6), 693–774.

Carroll, B. W., & Ostlie, D. A. (2017). *An introduction to modern astrophysics*. Cambridge University Press.

Caspar, M. (1993). *Kepler*. Dover.

DeWitt, R. (2011). *Worldviews: An introduction to the history and philosophy of science*. Wiley.

Dreyer, J. L. E. (1953). *A history of astronomy from Thales to Kepler*. Courier Corporation.

Fourier, J. B. J. (1822). *The analytical theory of heat, translated by Alexander freeman*. The University Press. translated from French.

Fraser, C. G. (2006). *The cosmos: A historical perspective*. Greenwood Publishing Group.

Freeth, T., Bitsakis, Y., Moussas, X., Seiradakis, J. H., Tselikas, A., Mangou, H., et al. (2006). Decoding the ancient Greek astronomical calculator known as the Antikythera mechanism. *Nature, 444*(7119), 587–591.

Guthrie, W. (1939). *Aristotle on the heavens*. Harvard University Press "Loeb Classical Library".

Hanson, N. R. (1960). The mathematical power of epicyclical astronomy. *Isis, 51*(2), 150–158.

Heath, T. (2004). *Aristarchus of Samos: The ancient Copernicus*. Courier Corporation.

Hetherington, N. S. (Ed.). (1993). *Cosmology: Historical, literary, philosophical, religous and scientific perspectives* (Vol. 1634). CRC Press.

Hirshfeld, A. W. (2002). *Parallax: The race to measure the cosmos*. Macmillan.

Kepler, J. (1596). Mysterium cosmographicum: The secret of the universe (trans. AM Duncan, introd. and commentary by EJ Aiton, 1981).

Kuhn, T. S. (1957). *The Copernican revolution: Planetary astronomy in the development of Western thought* (Vol. 16). Harvard University Press.

Lankford, J. (Ed.). (1997). *History of astronomy: An encyclopedia* (Vol. 1). Taylor & Francis.

Palter, R. (1970). An approach to the history of early astronomy. *Studies in History and Philosophy of Science Part A, 1*(2), 93–133.

Popper, K. (1958). Back to the pre-Socratics: The presidential address. In *Proceedings of the Aristotelian society* (Vol. 59, pp. 1–24). Aristotelian Society, Wiley.

Popper, K. (2012). *The world of Parmenides: Essays on the Presocratic enlightenment.* Routledge.

Roseveare, N. T. (1979). Leverrier to Einstein: A review of the mercury problem. *Vistas in Astronomy, 23*, 165–171.

Rovelli, C. (2011). The first scientist: Anaximander and his legacy.

Schoot, A. (2001). Kepler's search for form and proportion. Renaissance studies: Journal of the society for. *Renaissance Studies, 15*(1), 65–66.

Chapter 5
Conclusions

Throughout the analysis presented in this work, I have sought to extract meaningful conclusions that emerge from the convergence of multiple disciplines, each contributing distinct insights toward a shared objective: identifying the epistemological gaps that arise when scientific theories fail to account for empirical phenomena. To that end, it has been essential to first understand the theoretical debates surrounding emergence, and subsequently to explore how and under what conditions emergence can be quantitatively assessed. This, in turn, has led to a deeper examination of the ideas underpinning complexity metrics.

A central concern in this context is the nature of transitions between systemic states. A system's effective complexity cannot meaningfully increase—or change in any substantial way—if it remains confined to a single theoretical framework or within a static state space. To better understand how such transitions come about, this work introduced the notion of control parameters. These parameters describe the manner in which a system is coupled to its internal variables or to its external environment, and how such couplings govern the system's evolution over time.

From the perspective of an observer, high sensitivity to initial conditions or to parameter values signals predictive instability. In such cases, systems that are indistinguishable at the outset may evolve toward drastically different outcomes over time, diverging beyond the resolution of empirical detection—characteristic of chaotic dynamics.

Some control parameters—particularly those functioning as order parameters—can signal transitions to new forms of systemic organization. They may therefore serve as quantifiable indicators of weak emergence. Within this framework, bifurcation theory offers a formal starting point to understand how new structures and behaviors emerge when control parameters cross critical thresholds. The metaphor of a dynamic landscape, with basins of attraction shaped by non-equilibrium potentials and fluctuations, provides a useful way of visualizing and modeling such transitions.

M. Fuentes, *Complexity, Emergence and the Evolution of Scientific Theories: Towards a Predictive Epistemology*, SpringerBriefs in Philosophy,
https://doi.org/10.1007/978-3-031-99994-9_5

This work has proposed a series of conceptual tools that connect emergent properties to systemic complexity. Special emphasis has been placed on the epistemological implications of scientific theories: how they organize knowledge, frame empirical reality, and ultimately render themselves open to revision. Our treatment of emergence has insisted on recognizing the theory-ladenness of both observation and explanation.

A central concept here is that of parametric effective complexity. Defined as the Kolmogorov complexity of the regularities present in a system at a particular value of the control parameter, this function provides a powerful lens for assessing the epistemic adequacy of competing theories. It links theoretical models directly to the anomalies they struggle to account for, allowing us to interpret weakly emergent phenomena not merely as ontological novelties, but as transitions that expose the limitations of existing descriptive frameworks.

From this foundation, we introduced the concept of a parametric emergent property: a property is considered emergent at a particular value of the control parameter if the parametric complexity exhibits a discontinuous jump at that point. That is, if the system's effective complexity is significantly lower on one side of the parameter than on the other. This formulation not only sharpens the conceptual understanding of emergence but also connects it to the evolution of scientific theories and to the integration of new empirical findings within established frameworks.

This perspective also lends itself to a historical reading: across major scientific transitions, the selection of prevailing theories has often—implicitly or explicitly—reflected a preference for conceptual models that minimize parametric complexity. The accumulation of anomalies, the appearance of emergent self-organizing patterns, and the sheer volume of data made possible by technological innovation collectively place growing strain on existing theories. This process drives epistemic systems into a state of fragility, compelling theorists either to revise their conceptual tools or to adopt alternative frameworks that offer greater explanatory economy.

This brings us to a central insight of the book, which I refer to as predictive epistemology. As the complexity of a theory increases—reflected in the number of auxiliary hypotheses, ad hoc adjustments, or computational scaffolding required to sustain its predictive power—it may signal that the theory is approaching a state of epistemic exhaustion. In this light, rising complexity serves as a diagnostic signal: it tells us that the cognitive and practical cost of maintaining the theory may no longer justify its explanatory value.

This kind of epistemological inflation, where a theory becomes increasingly elaborate simply to remain coherent with accumulating data, closely parallels critical phenomena in dynamical systems. Just as a system may bifurcate when a control parameter exceeds a threshold, a scientific theory may undergo a paradigm shift when the internal resources it requires to remain viable surpass a threshold of complexity. At that moment, a new conceptual framework may emerge—one that is not only more consistent with the available data, but also more efficient in its representational and inferential capacities.

Importantly, such transitions should not be interpreted as failures of science, but rather as its adaptive evolution. Theoretical complexity is not merely a challenge to be overcome; it is also a driver of innovation. As our representations of the world grow more intricate, so too must the tools we develop to make sense of them. In this light, emergence becomes a catalyst for epistemic progress—a signal that new ways of thinking are needed, and that science remains a dynamic, evolving enterprise.

In conclusion, the rise of anomalies, the acceleration of data complexity, and the growing mismatch between theoretical models and observational realities all point to a critical insight: emergent phenomena and increasing effective complexity are not merely descriptive curiosities, but rather predictive indicators of theoretical transformation. The future of scientific understanding will hinge on our capacity not only to model these transitions but also to recognize the moments when our current models have reached their limits.

The manufacturer's authorised representative in the EU is Springer Nature Customer Service Centre GmbH, Europaplatz 3, 69115 Heidelberg, Germany. If you have any concerns regarding our products, please contact ProductSafety@springernature.com

Printed and bound by CPI Group (UK) Ltd, Croydon, CR0 4YY
01/10/2026
02226351-0009